CLIMB

The Journal of the International Space Elevator Consortium

(http://www.isec.org)

ACKNOWLEDGEMENTS

c⊔imb – The Journal of Space Elevator developments and technology

Official Publication of the International Space Elevator Consortium (ISEC)
http://www.isec.org

Publication & Review Committee

Editor-in-Chief
  Ted Semon
  President and Director, ISEC

Chief Technical Editor
  Ben Shelef
  Director & Technical Pillar lead, ISEC

Technical Review Committee
  Stephen Cohen, M. Eng.
    Editor, The Engineer's Pulse
  Blaise Gassend, PhD
    San Carlos, CA
  Benjamin H. Jarrell
    Jarrell & Doty, P.C.
  Martin Lades, PhD
    Nuremberg, Germany

ISEC Officers & Directors

Officers:
  President: Ted Semon
  Vice President: Peter Swan
  Secretary : Martin Lades
  Treasurer: Robert "Skip" Penny

Directors:
  Markus Klettner
  Martin Lades, PhD
  Bryan Laubscher, PhD
  Robert "Skip" Penny
  Ted Semon
  Ben Shelef
  Peter Swan, PhD

Pillar Leads:
  Technical: Ben Shelef
  Legal: Ben Jarrell
  Business: (Open)
  Outreach: Matt Gjertsen

c⊔imb (ISBN 978-0-557-29519-7) is published by the International Space Elevator Consortium (ISEC), 709A N Shoreline Blvd, Mountain View, CA, 94043, USA. Please direct all enquiries to CLIMB@isec.org.

Editorial Communications should be sent to the International Space Elevator Consortium, 709A N Shoreline Blvd, Mountain View, CA. 94043, USA or emailed to CLIMB@isec.org.

Subscription Rates: The Print version of c⊔imb is offered free, to all ISEC Members in good standing with membership level Professional or above. The electronic version of c⊔imb is offered free to all ISEC Members in good standing. Individual copies may be purchased from the ISEC store (located on the ISEC website; http://www.isec.org). For questions or more information, please email us at CLIMB@isec.org.

MISSION: Our Mission statement reads *"ISEC promotes the development, construction and operation of a space elevator as a revolutionary and efficient way to space for all humanity."* ISEC works towards this goal by publicizing and promoting the concept, promoting and encouraging research in technologies necessary to build a space elevator and promoting and encouraging research into all aspects of space elevator technology.

Disclaimer: The ideas and opinions expressed in c⊔imb do not necessarily reflect those of ISEC or the Editors of c⊔imb unless so stated. Articles contained in the Papers section of this journal have been reviewed and approved by the c⊔imb Technical Committee but again; do not necessarily reflect those of ISEC or the Editors of c⊔imb unless so stated.

Published by lulu.com.

CONTENTS

INTRODUCTION

PAPERS

OTHER READING

Юрий Арцутанов

(Yuri Artsutanov)

FOREWORD

When Arthur C. Clarke and I met in 1982, we still thought, as he put it, the space elevator would be built *50 years after people stopped laughing, in a couple of centuries or so*. So I am 'over the moon' to coin a phrase that the International Space Elevator Consortium *is bringing the reality of a Space Elevator much closer by* asking serious questions on how to solve the practical problems of: strength of the tether, powering the elevator, minimizing space junk and many more. I was the first to have written on this subject but as Clarke said it was an idea whose time has come. Development will accelerate. But, we are only at the stage James Watt was in the 1770s when he developed his steam engine. No one could foresee what applications it would spawn and we can only speculate on the future application of my space elevator idea. NASA is thinking of sending 20 people to Mars. They may build the first space elevator there, or on the moon, or ....between the satellites of Jupiter. As Jerome Pearson has suggested, the space elevator may play an important role in coping with climate change.

I was delighted *to meet Jerome Pearson again, and make the acquaintance of all the presenters and participants at the 2010 ISEC conference. Jerome Pearson's* proposal for reducing space junk is an interesting and vital contribution to bring the construction of a space elevator closer. ***The optimism of many of the other presenters was encouraging.*** The Annual space elevator conference in Seattle, which I had the privilege of attending, brought forth other questions. As we all know, asking the right questions is a vital step to finding answers.

Building a space elevator will require global cooperation. It has always been my hope that my idea might be a contribution to world peace, bringing all nations together to solve the legal, technical and social issues raised by realizing such a project. As I read in the last little while that rivalry in space is increasing I urge all to find means of working together. We all know of the possible dire consequences of unaccountable, uncontrolled projects in space.

This publication is proof of the vitality of the debate in many parts of the world. May readers be inspired by its content to help make extensive space travel a reality, spurred on by Tsialkovski's thought that mankind cannot stay in its cradle forever.

Yuri Artsutanov

August 2011

St Petersburg, Russia

PREFACE

Welcome to the first issue of CLIMB, a Journal devoted solely to the space elevator. Producing such a Journal has been a goal of the International Space Elevator Consortium (ISEC) since its inception and we are very happy that this day has arrived. CLIMB presents some of the best, peer-reviewed articles written on space elevator-related topics in the past several months as well as some additional papers we believe will be of interest to our readers. ISEC looks forward to producing additional issues of CLIMB on a regular basis in the future.

The pace of developments in the space elevator community continues to accelerate:

- Multiple space elevator competitions are now held on a yearly basis in the United States, Japan and, in 2011, for the first time, in Europe.
- Conferences devoted wholly or partially to the Space Elevator now also occur on a regular basis in the United States, Japan and Europe. At the 2010 US Space Elevator Conference, ISEC was privileged to sponsor the attendance of both Yuri Artsutanov and Jerome Pearson. Yuri is a Russian engineer and the true father of the modern day concept of the space elevator while Jerome is an American engineer and an independent co-inventor of the idea. For the 2011 Space Elevator Conference, ISEC sponsored the attendance of several experts in the carbon nanotube field. Expertise from researchers and academics such as these will be crucial in the development of a material strong enough to build a space elevator.
- ISEC has issued its first position paper, this on Space Debris Mitigation. Our second position paper, devoted to developments in area of carbon nanotube research, is scheduled for release in early 2012.
- ISEC continues to sponsor the Artsutanov and Pearson prizes; prizes intended to encourage research and development in the space elevator and related technologies. Though no prizes were awarded in 2010, two papers were judged worthy of Honorable Mention status and both of them have been included in this issue of CLIMB.
- The International Academy of Astronautics is now working on a document; *"Space Elevator – A Cultural Revolution"* with input from researchers around the world. In addition, ISEC now has its own team working on a similar project, along with tying it together with a business case scenario. The goal of the ISEC project is to show how building and operating a space elevator can be a profitable venture.
- The July, 2011 issue of National Geographic had a feature on the Space Elevator. In it, ISEC was credited with being a "Source". You can read this article online at: http://ngm.nationalgeographic.com/2011/07/visions-now-next#/next.

Let me close by encouraging you to consider joining ISEC. Our Mission Statement, *"ISEC promotes the development, construction and operation of a space elevator as a revolutionary and efficient way to space for all humanity"* says it all. We are doing everything we can to promote the idea of a space elevator and we need your help and your membership funds to continue to move forward. Please visit our website; http://www.isec.org to learn more about how you can become involved in this magnificent project, and don't forget to sign up for our free newsletter, keeping you up-to-date on space elevator-related developments.

Ted Semon, President – The International Space Elevator Consortium

PAPERS

AN UPDATED REVIEW OF NANOTECHNOLOGIES FOR THE SPACE ELEVATOR TETHER

G. Brambilla, E-mail:gb2@orc.soton.ac.uk
O.R.C., University of Southampton, Southampton SO17 1BJ, U.K.;

Abstract: This paper reviews the recent research on high strength materials in the nanowire form. After a brief overview of nanowires and nanotubes fabricated from different materials, this paper will provide an updated review of research on carbon nanotubes and related yarns.

1.   Introduction

Because of its extraordinary long length, a tether suitable for a space elevator requires 1) to stand extremely high stresses and 2) to be light (thus to have a relatively low density). If σ represents the material ultimate strength (defined as the maximum stress a material can withstand before breaking) and ρ its density, the specific ultimate strength σ_ρ (=σ/ρ) represents a good quality factor to evaluate the material suitability to be deployed in the future space elevator tether. In the SI system, the measurement unit of σ_ρ is $Pa/(kg/m^3) = (m/s)^2$; while in the textile industry σ_ρ is measured in N/Tex (1 N/Tex = 10^6 m^2/s^2), in this specific field the most commonly used unit is called Yuri (1 Yuri = 1 m^2/s^2). The strongest commercially available fibres (para-aramids, most commonly known as Kevlar® or Twaron®) have σ_ρ ~ 2 - 3 MYuri. A space elevator tether would ideally require at least σ_ρ > 20. The quest for a suitable material soon turned for an investigation of nanomaterials, which exhibit σ_ρ considerably larger than their macroscopic counterparts. Intuitively, this can be easily explained with Griffith's observation that cable strength is inversely proportional to the size of the largest crack on its surface [1]. Nanowires can only stand cracks that are a fraction of their diameter, thus they can exhibit strengths orders of magnitude larger than their bulk counterparts. In addition, at the nanoscale carbon presents a form (carbon nanotubes) which does not have a bulk counterpart. In the last decade four types of nanowires with σ_ρ > 5 MYuri have been reported in the literature.

2.   Nanowires and nanotubes

Fig. 1 summarizes the data on carbon nanotubes [2-5] (CNTs) silicon carbide [6] (SiC), silicon nitride [7] (Si_3N_4), and silica [8] (SiO_2). SiC and Si_3N_4 at their best can provide σ_ρ > 16 MYuri, while the greatest value reported for silica glass is smaller, σ_ρ ~ 12 MYuri. The great benefit of these nanowires relies in their possibility to be manufactured in extremely long lengths with minor changes to the current fabrication technology: silica glass allows for the prompt manufacture of km-long wires. Si_3N_4 can also be manufactured in relatively long lengths with the current technology, but the length of defect free Si_3N_4 single crystals has never been tested. CNTs have been identified as the ideal candidate because of their astonishing strength [1,2]: σ_ρ > 60 MYuri has been recorded for CNTs manufactured by chemical vapour deposition (CVD) with radii in the region of r ~ 50nm (Fig. 1).

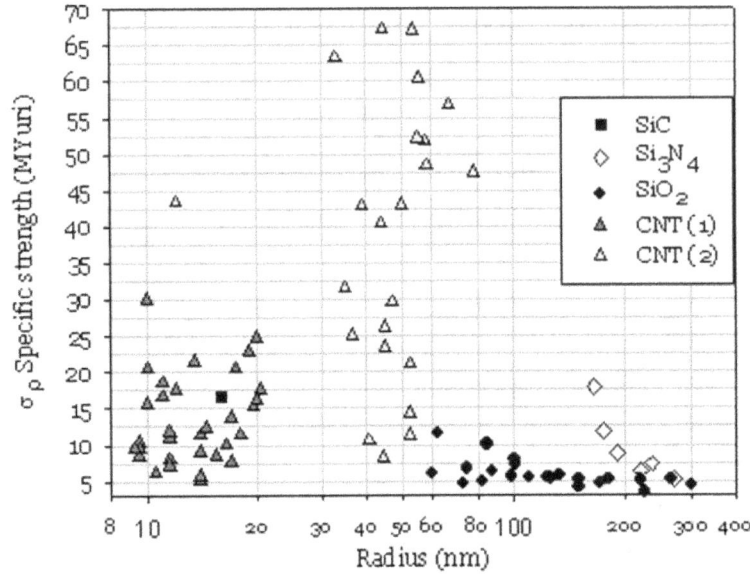

Fig. 1. Specific strength σ_ρ as a function of size for carbon nanotubes [2] (CNTs) and for silicon carbide [6] (SiC), silicon nitride [7] (Si₃N₄) and silica (SiO₂) nanowires [8]. CNTs have been fabricated by arc-discharge (group 1, older samples) and by chemical vapour deposition (CVD) (group 2, better samples).

Although this result is promising, calculations have shown that this value can be achieved for short-term loading. In fact, on the long term the excess elastic energy in a strained CNT is released through the spontaneous creation of topological defects [9] (like the Stone-Wales), the features of which depend on the CNT geometry and diameter.

Yet, the highest values of σ_ρ were measured on samples only few microns long [1,2] and long tethers are as strong as the weakest link. The presence of a single defect eventually decreases their ultimate strength: a single defect involving a missing atom in an otherwise perfect carbon nanotube 30000 km long would have a 20% decrease in the tensile strength with respect to that of a defect-free carbon nanotube [10]. Indeed, the fabrication of flawless wires has been proved very challenging.

3.   Ultralong CNTs & related issues

Since Iijima's publication in 1991 [11], the maximum length of single CNTs has continuously increased reaching fractions of a meter.

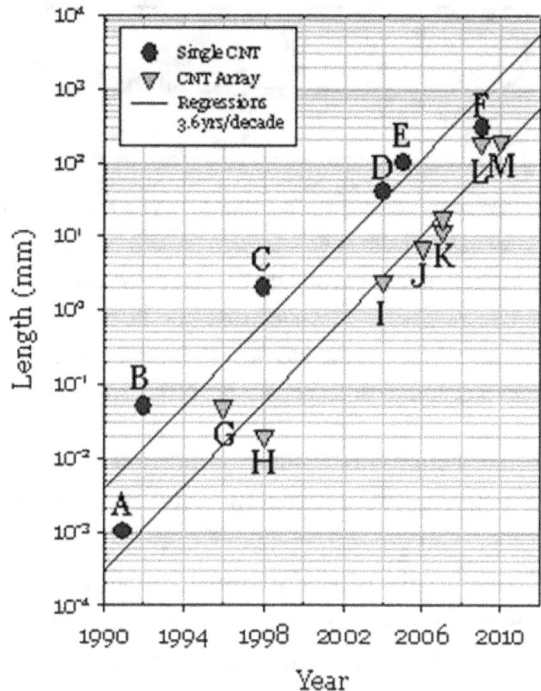

Fig. 2. Progress in carbon nanotube (CNT) maximum length for single CNT and CNT arrays. Regressions exhibit a similar slope, which predicts an increase of the CNT length by a factor of 10 every 3.6 years. Letters A-E and G-M relate to reference 11-15 and 16-22 respectively. F was reported on the internet but never formally published in the literature.

Fig. 2 reports the maximum CNT length reported in the literature in the last two decades. Since 2004 reports of CNTs longer than 1 mm have become common, with record lengths of 200 mm for a single CNT and for CNT bundles (manufactured in Beijing, China, at Tsinghua University [23], and reported in the literature in 2010).

If an equivalent of Moore's law can be established for the maximum length for CNT manufacture, the regression over last two decades would predict an order of magnitude increase in the length over 3.6 years, or a doubling approximately every year. This seems very promising news for the space elevator tether, since it predicts km-long CNTs being available by 2022. Still, two factors seem to strongly limit its practical realization.

Firstly, although CNTs have shown superb mechanical properties when tested in samples with micrometric lengths, σ_p has never been reported for "ultralong" CNTs. Yet, the uniformity of ultralong CNT electrical properties [21,22] could indicate a nearly flawless growth: in fact single defects in an otherwise perfect CNT significantly deteriorate the CNT electrical properties, especially at small CNT diameters. Yet, larger CNTs exhibit self-built tensile strains, which might affect their mechanical properties. This effect is likely caused by carbon nanodots on the CNT surface [24].

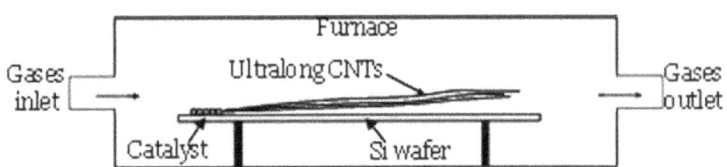

Fig. 3. Schematic of the process used to manufacture "ultralong" CNTs. A catalyst (Fe nanoparticles) is deposited onto a substrate (a silicon wafer slice) from FeCl3 solution (0.03 mol FeCl3 in 1 l ethanol): after ethanol evaporates, Fe-compounds are reduced in H2 atmosphere for 10 min at 900 0C. The temperature is increased to 1000 0C and a mixture of reagent gases (20 - 30 sccm CH4 + 40 - 80 sccm H2) is fed into the reactor. Small amounts of + H2O (0.01 - 1 vol. %) is added in the gas flow to keep the catalyst clean. Growth is stopped by switching off the reactant gases and cooling down the furnace in an inert (Argon) atmosphere.

Secondly, all so-called "ultralong" CNTs have been made using self-contained fabrication techniques, principally by chemical vapour deposition (CVD) using nano-catalysts (Fig. 3). This methodology [23] has been called kite growth mechanism, since most of the CNTs are floating in the gas flow and only sink down onto the substrate when the gas flow is stopped. This method requires the sample to be grown at high temperature in a very well controlled setting (normally a tube furnace) and the tube furnace size might limit the CNT maximum length [15]. Indeed, Fig. 2 seems to show a plateau in the length growth at few hundred millimetres.

4.   CNT growth

As far as the manufacture of unwound long CNTs is concerned, Fig. 4 shows that in the recent years fabrication has become faster and faster: at the beginning of 2010 [23] growth speeds in excess of 80 μm/s were achieved by using $CH_4 + H_2O$ as reagent gases.

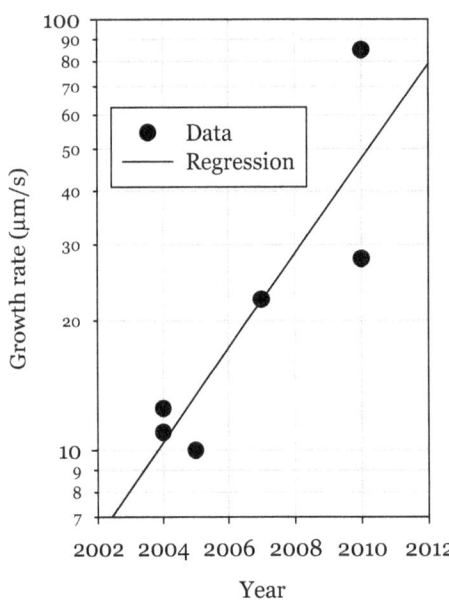

Fig. 4. Growth rate for CNT manufactured by CVD. Data points have been taken from refs. [14,15],[22,23],[27,28].

It is interesting to note that fast growth was achieved for multi-walled CNTs: indeed growth speeds of 80 - 90 μm/s were achieved with double- and triple- walled CNTs. Water addition to the gas mixture injected into the furnace has been shown to be fundamental for the quick manufacture of long CNTs: when water is not present among reagent gases, the catalyst can be deactivated due to a thin layer of carbon coating; water removes this coating and revitalizes the catalyst activity [25]. Despite initial concerns [15] regarding the limiting factor of the furnace hot zone length, the nature of the kite growth mechanism jointly with the revitalizing effect of water should potentially allow for the growth of unlimited lengths of CNTs; in fact, only the catalyst area has to be positioned within the high temperature zone. This would allow fabricating CNTs with conventional tube furnaces with only minor engineering to be performed to collect the long CNTs: a system winding CNTs on bobbins similar to that used for long yarns [26] can be envisaged.

5.   CNT yarns

The manufacture of CNT yarns has been implemented mainly using two-step techniques, which separate CNTs growth from yarn production. These techniques include, among the others, wet spinning in a solvent [29] or in a polymer solution [30] and, most recently, dry spinning [31].

Long yarns of CNTs have been tested [24] for strength and showed poor performance, but this result is more the consequence of the poor stress distribution within the CNT bundle and/or poor bonding between different CNTs than a sign of poor strength for a single CNT. Indeed, with better manufacturing techniques σ_p has increased from a fraction of MYuri ($\sigma \sim$ 0.7 GPa and $\rho \sim$ 800 kg/m^3) in 2004 [32] to 10 − 16 [33,34] MYuri ($\sigma \sim$ 3.3 GPa and $\rho \sim$ 200 kg/m^3) in 2007. Interestingly, it was observed that during stress tests the diameter shrank by as much as 10 %, suggesting that the maximum σ could be higher than the value reported in the literature. This effect was explained by a redistribution of stress within the yarn [35] (Fig. 5). CNTs have been shown to self-assemble into mesh-like structures [36].

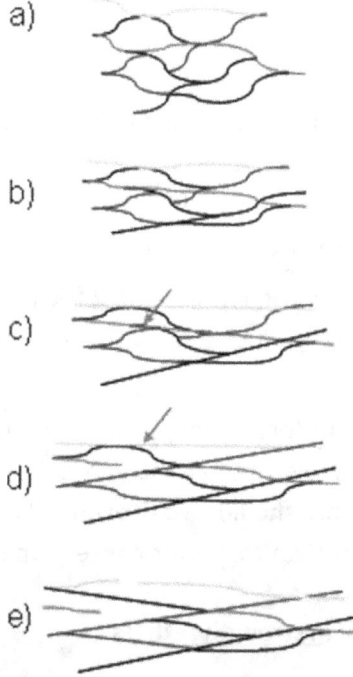

Fig. 5. Deformation process of a CNT yarn under macroscopic strain. For large strains overloaded CNTs fail (arrows) and strain is redistributed.

When strain is applied to the yarn, meshes are first deformed to longer and narrower structures (Fig. 5b) until some bundles are completely tightened (Fig. 5c); structures can then become overloaded (arrows in Fig. 5) and begin to fail (Fig. 5d,e); local stress is redistributed and at last only the strongest meshes carry the load before the final yarn rupture occurs. Raman spectroscopy has been shown as a possible investigative tool to measure in-situ, in-real-time CNT strain [37]: a 37.5 cm^{-1} shift per 1 % strain was observed in single CNTs. Analysis on yarns showed that in yarns CNT extension contributes only a small percentage to the macroscopic strain [36].

A process was developed in Cambridge (U.K.) to directly spin CNT yarns drawn from an aerogel sock [26]. Km long yarns have been manufactured at a rate of 20 m/min. Yet, σ_p decreased for increasing yarn lengths [38] from ~ 6 MYuri for gauge lengths in the region of 1mm to ~ 1 MYuri for gauge lengths of the order of 2 cm.

At the end of 2009 the quality of the CNT raw material [39] was thought to be a limitation to further improvement: by introducing a small amount of hydrogen during the growth, well-aligned CNT arrays have been obtained, considerably improving the yarn handling/fabrication with respect to case in air/oxygen atmosphere. Yarns longer than 40m with σ ~ 1.7 MYuri have been reported in 2010 [40]. Still, CNT alignment and uniform load distribution appears to be a major issue: in 2009, theoretical studies [41] predicted that σ_p has a dependence on the yarn length L (σ_L) and it is related to the CNT σ_p (σ_{CNT}) by:

$$\sigma_L = \sigma_{CNT} \sqrt{\frac{\left(\frac{\sigma_y}{\sigma_{CNT}}\right)^2 - 1}{\frac{l}{L} + 1} + 1} \qquad (01)$$

where 1 is a characteristic length (which corresponds to ~ 50μm for a defective CNT yarn with 10 % of distributed defects) and σ_y the yarn σp measured at very short lengths (L→0). A non-defective yarn with L~100,000 km is expected to have a 70% drop in strength with respect to the value of a single CNT, while a yarn with 10% distributed defects would have a 78% decrease over the same length [41].

The yarn σ_p has been shown to increase for increasing friction within the CNT bundle [42] and for post-fabrication exposure to microwaves, which decreases the number of defects [43], thus increases the strength.

Still, the fundamental requirement for a strong yarn is the uniform stress distribution amongst CNTs within the yarn; this requirement yields to an increased degree of orientation of CNTs along the longitudinal direction of the yarn. This should avoid overloading CNT bundles which considerably deteriorate the overall macroscopic mechanical performance of the yarn.

6.   Alternative possibilities for high strength tethers

A path to manufacture km-long tethers with σ_p > 20 MYuri could rely on the possibility to join different CNTs. Currently yarns are manufactured using CNTs which are relatively short (of the order of 1mm) and not connected at their extremities with other CNTs. In 2009 CNTs were connected with joints as strong as the CNT themselves [4] and, if parallel processing can be employed, long yarns could be manufactured from long CNTs.

Still, for enhanced yarn strength the critical issue is the uniform load sharing between different CNTs: up to date this issue has not been dealt with. Recently [44], a process to compact and better align CNTs along the yarn longitudinal direction has been demonstrated. Functionalization of CNT defect sites [45] can also be exploited: this could allow using functionalised sites to connect different CNTs, turning CNTs weak points (defects) in CNT yarn strength.

Another approach can rely on the combination of CNTs with other high-σ_p nanowires: silica nanowires can be an extremely good candidate because in principle they can be pulled over extremely long lengths using a process similar to that used in the manufacture of telecom optical fibres [46]. Indeed, if a telecom fibre is continuously feed into a furnace to manufacture a 1µm silica wire, a single bobbin (usually containing a single, 10 - 20 km-long strand) of telecom fibre would provide an interrupted single cable 150,000 - 300,000 km long. This could represent a backbone on which CNTs are fixed, possibly on a carbon layer [47] acting as a buffer. A technique similar to the Outside Vapour Deposition [48] (OVD), developed by Corning Inc. to manufacture optical fibre performs, can be envisaged to fabricate such a glass/CNT composite. By changing the ratio between carbon and glass it should be possible to achieve high σ_p over long lengths. The carbon phase needed to achieve high strength is still limited to CNT or graphene (which has $\sigma \sim 130\pm10$ GPa [49]): in fact, the "third-best" choice (tetrahedral amorphous carbon also called diamond-like carbon (DLC)) has a strength $\sigma \sim 25$ Gpa [50] and, although it is already used to provide uniform uninterrupted glass coatings, it exhibits these enhanced properties only over short lengths [51]. Still DLC can be used as embedding material for CNTs and to join their extremities, providing a technique capable to spread the load among CNTs.

Finally, the use of CNTs as reinforcing fibre in conventional matrixes has been widely investigated [52]; although the matrix overall strength is considerably increased, recorded strength values are orders of magnitude smaller than those achieved for CNTs, making this approach still without prospect for any application as material of choice for the space elevator tether.

References

[1] Griffith, A.A., Phil. Trans. Royal Soc. London 221A, 163 (1921).

[2] Barber, A. H., et al. Comp. Sci. Technol. 65, 2380 (2005).

[3] Yu, M. F., et al., Science 287, 637 (2000).

[4] Cui, Q., et al., Small 5, 1246 (2009).

[5] Wang, M. S., et al., Adv. Funct. Mat. 15, 1825 (2005).

[6] Wong, S. S., et al., Science 277, 1971 (1997).

[7] Iwanaga, H., et al., J. Am. Ceram. Soc. 81, 773 (1998).

[8] Brambilla, G., et al., Nano Lett. 9, 831 (2009).

[9] Zhao, Q., et al., Phys. Rev. B 65, 144105 (2002).

[10] Pugno, N. M., and Ruoff, R. S., Philos. Mag. 84, 2829 (2004).

[11] Iijima, S., Nature 354, 56 (1991).

[12] Ebbesen, T. W., and Ajayan, P. M., Nature 358, 220 (1992).

[13] Pan, Z. W., et al., Nature 394, 631 (1998).

[14] Zheng, L. X., et al., Nat. Mat. 3, 673 (2004).

[15] Hong, B. H., et al., J. Am. Chem. Soc. 127(44), 15336 (2005).

[16] Li, W. Z., et al., Science 274 (5293), 1701 (1996).

[17] Kong, J., et al., Nature 395, 878 (1998).

[18] Hata, K,.et al., Science 306(5700), 1362 (2004).

[19] http://www.uc.edu/News/NR.aspx?ID=4811.

[20] http://www.uc.edu/News/NR.aspx?ID=5700.

[21] Wang, X.,et al., Nano Lett. 9 (9), 3137, (2009).

[22] Wen, Q., et al., Adv. Mater. 22(16), 1867, (2010).

[23] Wen, Q., et al., Chem. Mater. 22, 1294 (2010).

[24] Gao, P., et al., ACS Nano, 4(2), 992 (2010).

[25] Yamada, T., et al., Nano Lett. 8(12), 4288 (2008).

[26] Li, Y-L, et al., Science 304, 276 (2004).

[27] Huang, S., et al., Nano Lett. 4(6), 1025 (2004).

[28] Yao, Y.G., et al., J. Phys. Chem. C 111(24), 8407 (2007).

[29] Ericson, L.M., et al., Science 305, 1447 (2004).

[30] Vigolo, B., et al., Science 290(5495), 1331 (2000).

[31] Miao, M., et al., Carbon 48(10), 2802 (2010).

[32] Zhang, M., et al., Science 306, 1358 (2004).

[33] Koziol, K., at al., Science 318(5858), 1892 (2007).

[34] Zhang, X., et al., Adv. Mater. 19, 4198 (2007).

[35] Ma, W. J., et al., Nano Lett., 7, 2307 (2007).

[36] Ma, W. J., et al., Adv. Mat., 21, 603 (2009).

[37] Cronin, S. B., et al., Phys. Rev. B, 72, 8 (2005).

[38] Stano, K.L., et al., Int. J. Mater. Form. 1(2), 59 (2008).

[39] Zheng, L., et al., Small 6(1), 132 (2009).

[40] Liu, K., et al., Nanotech. 21(4), 045708 (2010).

[41] Pugno, N.M., Meas. Sci. Technol. 20, 084028 (2009).

[42] Zhang, X., et al., ACS Nano, 4 (1), 312, (2010).

[43] Lin, W., et al., ACS Nano, 4(3), 1716 (2010).

[44] Tang G., et al., Small, 6(14), 1888 (2010).

[45] Sahoo, N.G., et al., Progr. Polymer Sci. 35(7), 837 (2010).

[46] Vukovic, N., et al., Phot. Technol. Lett. 20(14), 1264 (2008).

[47] David, M.M., et al., patent US5,609,948.

[48] Berkey, G.E., et al., conf. on Opt. Fiber Comm., Tech. Digest, paper ThCC5 (1982).

[49] Lee, C., et al., Science 321, 385 (2008).

[50] Espinosa, H.D., et al., Appl. Phys. Lett. 89, 073111 (2006).

[51] Jonnalagadda, K. N., et al., J. Nanomat. 2009, 204281, (2009).

[52] Spitalsky, Z., et al., Progr. Polymer Sci. 35(3), 357 (2010).

OPTIMAL DESIGN OF THE SPACE ELEVATOR TETHER

Ambartsumian S.A. (samb@sci.am),
Belubekyan M.V. (mbelubekyan@yahoo.com),
Ghazaryan K.B. (ghkarren@gmail.com)
Institute of Mechanics, National Academy of Sciences, Yerevan, Armenia

Abstract. Assuming that a proper material (carbon nanotubes) is available, we examine the optimal design problems of a thin vertical tether stretched from the Earth surface to the geostationary orbit and a counter-mass far out in space. Different optimal problems are proposed to reduce the tether's strength and length. On the basis of analysis of suggested optimization problems new qualitative and quantitative results are obtained for reducing the tether's maximal strength and length (volume).

1.   Introduction

The theoretical possibility of constructing a tower to connect a geostationary satellite to the ground was first introduced in [1], where the problems of buckling, strength and dynamic stability of a tower construction were studied.

In recent years many papers have been published on different topics related to the engineering, technical, environmental impact and application problems of the space elevator construction. An overview and developments of the space elevator concept, detailed design for the construction and operation of a space elevator are given [2,3].

An alternative concept of the lunar space elevator was invented in [4,5]. The concept combines lunar space elevators with solar-powered robotic climbing vehicles, a system for lunar resource recovery, and orbit transfer space vehicles to carry the lunar material into the high Earth orbit. The main advantage of the lunar space elevator from the Earth space elevator is that the lunar space elevator can be constructed with the existing contemporary high-strength, lightweight materials, like Kevlar or Spectra [5] while the Earth space elevator will require carbon nanotubes [3].

In [6] two dynamic models of a tethered space elevator system have been derived based on a lumped mass approximation. One model represents the elevator dynamically, where its motion is governed by applied forces, whereas the second model represents the elevator's motion kinematically by adjusting the tether lengths adjacent to the elevator.

The space elevator behavior and stability in space affected by the attractions of the detailed shape of the Earth, the Sun and the Moon and by presence of space debris have been discussed in [7].

The problem of the existence and stability of the radial relative equilibrium of a tapered string on a circular geosynchronous trajectory around the Earth is studied in [8].

The paper [9] is devoted to the problem of the elastic oscillation of the space elevator tether, where for a small motion longtitudinal and transversal modes of oscillation are analysed.

Authors of the paper [10] attempt to develop a realistic and yet simple planar dynamic model describing the interaction of a space elevator two components - tether and climber.

In the framework of the space elevator concept a theoretical possibility has been examined for superconducting elastic cable transmitting an electrical current from space in [11].

In [12] the role of the thermodynamically unavoidable atomistic defects with different size and shape is quantified on brittle fracture, fatigue and elasticity, for nanotubes and nanotube bundles. Elastic plasticity, rough cracks, finite domains and size effects are also discussed. The results are compared with atomistic simulations and nanotensile tests of carbon nanotubes.

In [13] an overview of the radiation belts in terms relevant to space elevator studies is given. The expected radiation doses as computed and the required level of shielding is evaluated. Models of passive shielding using aluminum and active shielding using magnetic fields are discussed.

The strength problem was first solved by J. Pearson [1], who suggested overcoming the strength requirements by tapering tether cross-section (with very large non reasonable taper ratio) as an exponential function from a maximum at the geostationary point to a minimum at the ends.

In continuation of our previous studies [14,15] assuming that a proper material (carbon nanotubes) is available we propose here a new optimal strength problem based on the conception of a cross-section tapering. Considering the volume of the ribbon as a function of the optimal design and the strength as a function of the objective, we show that taking the tether stress uniform at the neighborhood of the geostationary point, it is possible to minimize the strength and length of the tether for reasonable taper ratios. The optimal size of equal strength (uniform stress) zone is determined. In the second optimal project a piecewise homogeneous tether model is proposed. The tether under consideration has different piecewise constant cross-sections, thickening at the neighborhood of the geostationary point. It is shown that by means of the appropriate choice of the tether geometry one can minimize the maximum stress and length of the tether. The comparison of two optimization problems is carried out. Another optimal problem is solved for an inhomogeneous tether when the density of the tether material is a function of its length. It is shown that the optimal project is a compound tether made from two homogeneous materials of different densities fastened at the geostationary point, whereas the density of the tether upper part must be bigger than the density of the lower part.

2. Problem Statement and Basic Equations

Consider a very long elastic tether of length l anchored on the Earth equatorial point with a counter-mass m_0 far out in space. The tether is subjected to the action of the Earth gravity inward force F_1, defined by the Newton gravity law and centrifugal outward force due to the Earth's daily spinning.

$$F_1(\gamma) = \frac{\rho g_0 R_0^2}{\left(R_0 + \gamma\right)^2}, \quad F_2(\gamma) = -\rho \omega^2 \left(R_0 + \gamma\right), \qquad (1)$$

where γ is a coordinate along the tether length (altitude) counted from the Earth surface, ρ is the bulk density of the tether material, $R_0 \approx 6378 km$ is the Earth's equatorial radius, $g_0 = 980 sm \cdot sec^{-2}$ is the gravity force acceleration on the Earth's surface, $\omega = 2\pi/T$ is the circular spinning frequency of the Earth, $T = 86146 \sec$ is the period of the Earth's spinning.

In this paper we take into account only two mechanical forces given by formulae (1). Apart from these basic mechanical loads there are a number of other forces acting on the space elevator to balance the resulting forces e.g. the interaction forces of the Sun, the Moon, other space objects, space debris, wind, meteors, lighting, electromagnetic fields etc. Some estimations of such forces are given in [2,3]. The influence of space debris on space flights and space constructions is discussed in [13,16] where practical suggestions and concepts of the shield design are proposed. The environmental hazards are very important and their study in the space elevator design deserves a special attention.

The equation of the tether equilibrium under the action of mechanical forces (1) can be written as

$$\frac{d\left[S(\gamma)\tilde{\sigma}(\gamma)\right]}{d\gamma} + S(\gamma)\left[\frac{\rho g_0 R_0^2}{\left(R_0+\gamma\right)^2} - \rho\omega^2\left(R_0+\gamma\right)\right] = 0,$$

where $S(\gamma)$ is the cross-section area, $\tilde{\sigma}(\gamma)$ is the stress of the tether. Using the following dimensionless notations

$$x=\gamma/R_0 , \quad \alpha=\omega^2 R_0/g_0 \approx 1/288 ,$$

$$L=l/R_0 , \sigma=\tilde{\sigma}/\rho R_0 g_0$$

we can write the above equation in the following dimensionless form

$$\frac{d\left[S(x)\sigma(x)\right]}{dx} - S(x)g(x) = 0 \qquad (2)$$

where

$$g(x) = \left[\alpha(1+x) - \frac{1}{(1+x)^2}\right].$$

We consider Eq. (2) with the following boundary condition for the tether attached to the counterweight mass m_0 at the outward end $x=L$

$$\sigma(L) = \frac{m_0 g(L)L}{m} ,$$

where $m=\rho L R_0 S_0$ is the mass of the tether.

In order to avoid a tether buckle the tether should be under tension over the whole length $\sigma(x)\geq 0$. To ensure this we take $\sigma(0)=0$. This condition defines the minimum length $L_0\sqrt{2}$ for the tensile tether.

The stress function satisfying the boundary condition at the outward end can be found as

$$\sigma(x) = \frac{(L-x)\left[(1+L)(1+x)(2+L+x)-576\right]}{576(1+L)(1+x)} + \frac{m_0 L}{m}\left[\frac{(1+L)}{288} - \frac{1}{(1+L)^2}\right]$$

For practically realistic cases the counterweight mass decreases values for the minimal length and stress negligibly if $m >> m_0$.

Table 1.

Values for the minimum length and stress depending on the counter mass ratio

m/m_0	100	50	20	10
L_0	22.3	22.07	21.0	20.48
σ_0	0.76	0.74	0.66	0.63

J. Pearson [1] has suggested overcoming the strength requirements by tapering the tether cross-section (with very large non reasonable taper ratio) as an exponential function from a maximum at the geostationary point to a minimum at the ends. This uniform-stress model makes it possible to design a tether which theoretically could be built of any material by simply using a rather large taper ratio. However in practice, the taper ratio required for normal structural materials is very large resulting in an enormous area at the synchronous point for a reasonable area at the ends.

3. Uniform stressed tether

Using the concept of a cross section tapering here a model is proposed assuming a constant stress only in the neighborhood of the geostationary orbit. This model allows reducing both the tether stress and length.

Setting $\sigma(x) = const$ in (1) results in the following equation defining the cable's cross-section area in the interval $x_1 < x < x_2$

$$\sigma_0 \frac{d[S(x)]}{dx} - g(x)S(x) = 0 \tag{3}$$

Taking into account that the functions $S(x)$ and $\sigma(x)$ are continuous in the whole interval $0 < x < L$ and integrating equation.(3) gives

$$S(x) = S_0 \exp\left[\frac{f(x) - f(x_2)}{\tilde{\sigma}_0}\right], x \in (x_1, x_2), \tag{4}$$

where S_0 is the cross section area of the upper end and

$$f(x) = \int_0^x g(x)dx = \frac{x[574 - x(3+x)]}{576(1+x)} \tag{5}$$

Note that the function $f(x)$ defines the stress shape for a non-tapered tether when the counter-mass load is negligible. The function $f(x)$ has the following properties

$$f(x) \geq 0, \quad \max f(x) = f(x_0) = 0.7746; \quad F(0) = F(L_0) = 0$$

where x_0 is the dimensionless coordinate of the geosynchronous orbit and $L_0 = 22.053$ is the minimum length of the non tapered tensile tether.

In the interval $0 < x < L$ we have the following shape for the cross-section function

$$S(x) = S_0 \exp\left[\frac{f(x_1) - f(x_2)}{\sigma_0}\right], \quad 0 < x < x_1,$$

$$S(x) = S_0 \exp\left[\frac{f(x) - f(x_2)}{\sigma_0}\right], \quad x_1 < x < x_2, \quad S(x) = S_0, \quad x_2 < x < L. \tag{6}$$

here $x_1 \leq x_0 \leq x_2$.

Integrating equation (2) gives the following solutions for the stress function

$$\sigma(x) = f(x), \quad 0 < x < x_1,$$

$$\sigma(x) = \sigma_0 = f(x_1), \quad x_1 < x < x_2,$$

$$\sigma(x) = f(x) - f(L), \quad x_2 < x < L.$$

Note that here the small corrections caused by the attached mass m_0 applied at $x = L$ are not taken into account.

From the continuity condition of the stress function it follows that

$$f(x_2) - f(x_1) = f(L). \tag{7}$$

Equation (7) defines the minimal length of the tether.

The location of the interval (x_1, x_2), where a stress is constant, is not defined uniquely. There is an infinite set of points x_1 and x_2 corresponding to each prescribed value σ_0.

Figures 1-4 illustrate this for $\sigma_0 = 0.65$ where different profiles for functions $\sigma(x)$ and $S(x)/S_0$ are presented. In Fig. 1-5 the plot of the cross-section function is presented as a continuous curve and the plot of the stress function is presented as a dashed curve.

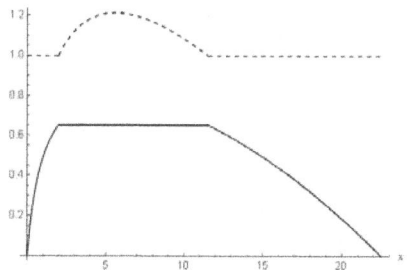

Fig 1. $x_1 = 1.972, x_2 = 11.512, K = 1.211, V = 23.835, L = 22.505$.

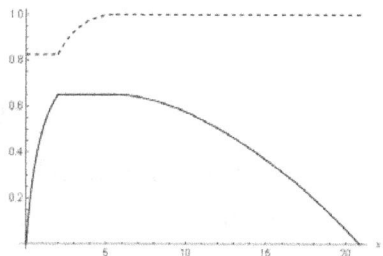

Fig 2. $x_1 = 1.972, x_2 = 5.602, L = 20.883, K = 1.211, V_0 = 20.356$

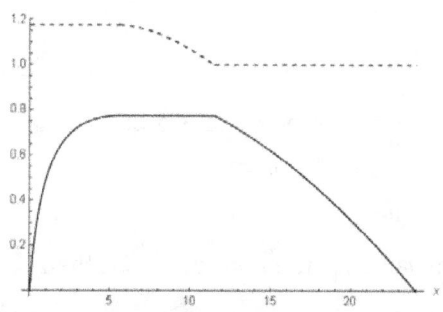

Fig 3. $K = 1.211, V = 25.641, L = 24.014, x_1 = 5.602, x_2 = 11.512$.

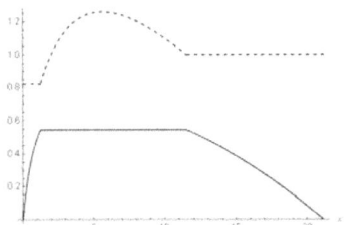

Fig 4. $K = 1.532, V = 22.45, \ L = 21.117, x_1 = 1.22, x_2 = 11.512$

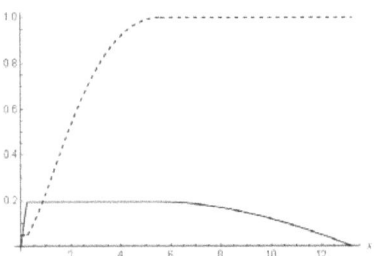

Fig 5. Data are given in Table 1 (next to the last column).

The plots in Fig. 1-5 show how the stress and the cross-section functions vary along the tether length (V is the volume of the tether, K is the taper ratio) and

$$V = \frac{\displaystyle\int_0^L S(x)dx}{S_0}, \qquad K = \max S(x)/\min S(x)).$$

In order to choose the optimal project we should find the location for the points x_1, x_2 which ensures the minimum for the tether volume $V(x_1, L)$ and the taper ratio K for a given constant value of the stress σ_0.

Thus we can state that the location of the point x_1 defining the optimal project will be determined as a minimum positive root of the following equation

$$f(x) - \sigma_0 = 0 \tag{8}$$

The second point will coincide with the dimensionless coordinate of the geosynchronous orbit $x_2 = x_0 = 5.602$. The profiles of the optimal projects are presented in Fig.2 and Fig.5.

Examples of optimal projects are presented in Table 2 for several values of the constant stress σ_0 in the interval $x \in (x_1, x_0)$, where $\sigma_{0*} = 0.774$ is the stress maximum value for the non-tapered tether.

Table 2

$\sigma_0 \times$	0.9	0.8	0.7	0.6	0.5	0.25	0.1
σ_0	0.69	0,61	0.54	0.46	0.38	0.194	0.07
x_1	2.53	1.71	1.21	0.88	0.63	0,212	0.084
L_0	21.51	20.46	19.36	18.19	16.93	13.20	10.11
V_0	21.14	19.84	18.52	17.13	14.03	11.17	7.15
K_0	1.11	1.28	1.535	1.94	4.48	20.08	8102

The data in Table 2 show that we can significantly reduce the stress maximum value (up to four times) and the volume (up to two times) for a realistically taper ratio not exceeding 20-25. Further reduction leads to a bigger increase in the taper ratio making the projects unrealizable.

4.   Stepped tether

Consider now a tether with a piecewise constant cross-section area $S(x)$

$$(S_1 < S_2)\, S(x) = \begin{cases} S_1, & 0 < x < x_1, \\ S_2, & x_1 < x < x_2, \\ S_3, & x_2 < x < L, \end{cases}$$

where $x_1 < x_0 < x_2$.

Taking into account that the stress resultants $T(x) = S(x)\sigma(x)$ are continuous at x_1, x_2
$$S_1\sigma_1(x_1) = S_2\sigma_2(l_1), \quad S_2\sigma_2(x_2) = S_3\sigma_3(x_2)$$
and integrating equation (3) gives the following solutions for the stress function

$$\sigma(x) = \begin{cases} \sigma_1 = f(x); \ 0 < x < x_1 \\ \sigma_2 = f(x) - (1-k_1)f(x_1); x_1 < x < x_2 \\ \sigma_3 = f(x); x_2 < x < L \end{cases}$$

where $k_1 = S_1/S_2 < 1$. From the continuity condition of the stress the resultants at x_2 it follows that

$$k_3 f(L) - (1-k_1)f(x_1) + (1-k_3)f(x_2) = 0$$

where $k_3 = S_3/S_2 < 1$ This condition defines the minimum length of the tether.

In order to choose the optimal project we should find a location for the points x_1, x_2 which for the prescribed values of the taper ratios k_1, k_3 ensures the minimum for the stress value σ_0 and the tether volume V

$$\sigma_0 = f(x_0) - (1-k_1)f(x_1).$$

$$V = S_2[x_1 + k_1(x_2 - x_1) + k_3(L - x_2)].$$

The location of the points x_1 and x_2 defining the optimal project can be found from the following conditions

$$\sigma_1(x_1) = \sigma_2(x_0) = \sigma_3(x_2) = \sigma_0 \tag{9}$$

Using (9) we can rewrite the stress function profile as

$$\sigma_1 = f(x), \quad 0 < x < x_1,$$
$$\sigma_2 = f(x) - (1 - k_1)\sigma_0, \quad x_1 < x < x_2,$$
$$\sigma_3 = f(x) - (k_3 - k_1)\sigma_0, \quad 0 < x < x_1,$$

where

$$\sigma_0 = \frac{f(x_0)}{2 - k_1}$$

is the maximum value of the stress function.

The location of the points x_1 and x_2 defining the optimal project are determined from the following equations

$$f(x_1) = \sigma_0 \tag{10}$$

$$f(x_2) = (1 + k_3 - k_1)\sigma_0 \tag{11}$$

Equation (10) has positive roots when $k_1 \leq 1$. The minimum positive root of equation (10) corresponds to the point x_1. The condition for a minimum length can be written as

$$f(L) = (k_3 - k_1)\sigma_0. \tag{12}$$

Equation (12) has positive roots only for $k_3 \leq 1$, $(k_1 \leq 1)$. On the other hand it is easy to verify that the value of a positive root of equation (12) decreases with the growth of k_3. Combining these two facts we should take $k_3 = 1$ in order to obtain the minimum value for L. When $k_3 = 1$, equation (12) has one positive root equal to the dimensionless coordinate of the geosynchronous orbit $x_2 = x_0 = 5.602$. This implies that the volume of the stepped tether can be calculated by the formula

$$V = S_2[L - (1 - k_1)x_1].$$

Thus we can conclude that for the optimal tether consisting of the uniform parts we should take $S_1 < S_2$; $S_3 = S_2$; $(k_3 = 1)$. In this case the maximum stress and the length of the tether will be less than the maximum stress and the length of the non-tapered tether.

Table 3

Examples of optimal projects presented for several values of the taper ratio ($V_0 = V / S_2$).

k_1	0.9	0.8	0.7	0.6	0.3	0.1	0.05
σ_0	0.70	0.64	0.59	0.55	0.45	0.40	0.39
x_1	2.64	1.92	1.53	1.27	0.85	0.69	0.666
V_0	21.33	20.43	19.67	19.01	17.45	16.657	16.57
L	21.60	20,82	20.13	19.52	18.05	17.27	17.10

The graph of the stress function profile for the stepped tether is presented in Fig.6 for the optimal project corresponding to the taper ratio $k_1 = 0.1$ (data next to the last column in Table 3).

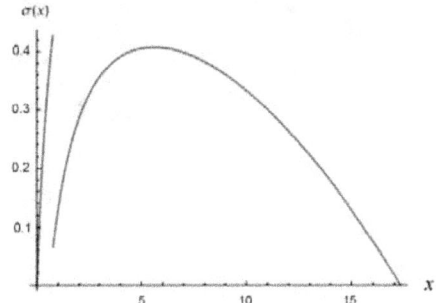

Fig.6

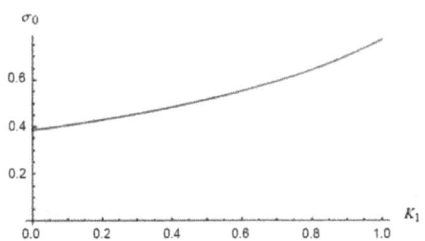

Fig.7.

Figure 7 also shows the behavior of σ_0 and the taper ratio $k_1 = 1/K$ for the optimal projects.

According to the data in Table 3 the stress value for a stepped tether can be reduced only two times compared to the uniform-stressed tether. The comparison of the underlined data in Table 2 and Table 3 shows that for the same taper ratio at geosynchronous orbit the decrease of the stress maximum value, length and volume are more considerable in the case of a uniform-stressed tether. But the stepped tether has an important advantage that its realization is simpler and easier from the technical point of view.

5. Tether made from a non-homogeneous material

Consider a tether made from a non- homogeneous material assuming that the density of the ribbon material is a function of its length. For this ribbon we would like to formulate the following optimal problem, the solution of which is important from an applications point of view. Considering the density $\rho(x)$ of the tether as a function of the design and the stress as a function of the objective, we should find the optimal solution for which

$$\max_{x} \sigma(x) \to \min_{\rho(x)}, \quad \rho_{01} < \rho(x) < \rho_{02}$$

under restriction $\sigma(0) = 0$.

The following solution for the stress function follows from equation (2)

$$\tilde{\sigma}(x) = R_0 g_0 \int_L^x \rho(\eta) g(\eta) d\eta$$

It is obvious that

$$\max_x \tilde{\sigma}(x) = R_0 g_0 \int_L^{x_0} \rho(\eta) g(\eta) d\eta$$

Taking into account the restriction $\sigma(0) = 0$ we have

$$\max_x \tilde{\sigma}(x) = R_0 g_0 \int_0^{x_0} \rho(\eta) g(\eta) d\eta \tag{13}$$

It follows from equation (13) that the function

$$\rho(x) = \begin{cases} \rho_1; x \in [0, x_0]; \\ \rho_2; x \in [x_0, \tilde{L}_0] \end{cases}$$

is the optimal solution of the problem.

The minimum value of $\max_x \sigma(x)$ can be defined as

$$\tilde{\sigma}_0 = 0.7746 \rho_1 R_0 g_0,$$

while the minimum length of the tensile tether $\tilde{L}_0$ can be defined from the following integral equation

$$\rho_2 \int_{\tilde{L}_0}^{x_0} g(\eta) d\eta = 0.774 \rho_1$$

or

$$0.774(1 - \beta) = \int_0^{L_0} g(\eta) d\eta \tag{14}$$

$$\beta = \rho_1 / \rho_2 .$$

Figure 8 shows the minimal length $\tilde{L}_0$ of the tether depending on of parameter β

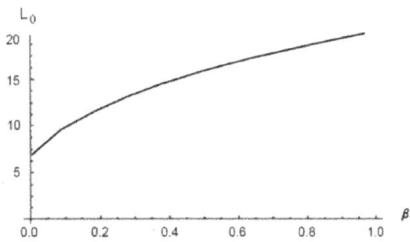

Fig. 8

Thus if we have two materials with sufficient strength limits and different densities ($\rho_1 < \rho_2$) the tether can be constructed in the following way. The lower part of the tether before the geosynchronous orbit must be constructed with the first material (ρ_1), the upper part with the second material (ρ_2). The more the difference between densities the less the whole tether's length $\tilde{L}_0$. If $\rho_1 \ll \rho_2$ the tether upper end (length) becomes closer to the geosynchronous orbit.

6.   Conclusions

In this paper we have presented three optimal projects for decreasing the maximum stress value, length and volume of a space elevator tether for a realistic taper ratio of the cross-section area in the neighborhood of the geostationary point. The corresponding optimal problem is stated and solved for the tether with a constant stress in the neighborhood of the geostationary orbit. The same optimal problem is solved for the stepped tether with a piecewise constant cross section. For both cases, the graphs illustrating the shapes of the cross-section areas and stress profiles for the optimal projects are provided. The numerical data describing how much we can decrease the maximum stress value, length and volume of space elevator tether as a function of the taper ratio are presented. The optimal problem is studied for a tether made from two materials with different densities and fastened together at the geosynchronous orbit. It is shown that a big difference between material densities leads to a considerable decrease in the tether length if the density of the tether's upper part is bigger than the density of the tether's lower part. The results are important from an applications point of view, since they allow us to construct a realistically tapered tether, non-homogeneous tether with a less length, volume and a value of maximum stress than those of an homogeneous and non tapered tether.

References

[1]   J. Pearson, The orbital tower: a spacecraft launcher using the Earth's rotational energy, Acta Astronautica. 1975, Vol. 2. pp. 785-799.

[2]   B. C. Edwards, Design and Deployment of a Space Elevator. 2000, Acta Astronautica. 47, pp.735-744.

[3]   B. C. Edwards, E. A. Westling., The Space Elevator, Houston, USA, 2002, p.280

[4]   J. Pearson, Anchored Lunar Satellites for Columnar Transportation and Communication, Journal of the Astronautical Sciences, Vol. XXVII, No. 1, pp. 39-62, Jan/Mar 1979.

[5]   J. Pearson, E. Levin, J. Oldson, and H. Wykes, The Lunar Space Elevator, IAC-04-IAA.3.8.3.07, 55th International Astronautical Congress, Vancouver, Canada, October 4-8 2004.

[6]   P. Williams, Dynamic multibody modeling for tethered space elevators, Acta Astronautica, Volume 65, 2009, pp. 399-422.

[7]   P. Lubos, Space Elevator: Stability, Acta Astronautica, v. 62, 2008, pp .514-520.

[8]   N. Pugno, M. Schwarzbart, A. Steindl, Hans Troger, On the stability of the track of the space elevator, Acta Astronautica, 2009, pp. 524-537.

[9]   Stephen S. Cohen, Arun K. Misra, Elastic of the space elevator ribbon, Journal of guidance, control and dynamics, v.30, 6, 2007, pp.1711-1717.

[10] Stephen S. Cohen, Arun K. Misra, The effect of climber transit on the space elevator dynamics, Acta Astronautica, March-April 2009, pp. 538-553.

[11] S. A. Ambartsumian, M. V. Belubekyan, K. B. Ghazaryan, Stability of superconducting cable used for transportation of electrical current from space, Acta Astronautica, Vol: 66, 2010, pp. 563-566.

[12] N. M. Pugno, The role of defects in the design of space elevator cable: From nanotube to megatube, Acta Materialia, v. 55, Issue 15, September 2007, pp. 5269-5279.

[13] A. M. Jorgensen, S.E. Patamia, B. Gassend, Passive radiation shielding considerations for the proposed space elevator, Acta Astronautica, Volume 60, 2007, pp. 198-209.

[14] S. A. Ambartsumian, M. V. Belubekyan, K. B. Ghazaryan, Some Aspects of Space Elevator Ribbon Stress and Length Reduction, Reports of Armenian National Academy of Sciences,107, 4, 2007, pp. 345-352.

[15] S. A. Ambartsumian, M. V. Belubekyan, K. B. Ghazaryan., and V. Tc. Gnuni, On design problem of space elevator cable. Reports of Armenian National Academy of Sciences, 104, 3, 2004, ISSN 0321-1339, pp. 180-186.

[16] N. N. Smirnov, Space Debris-Hazard Evaluation and Mitigation, Earth Space Institute Series, Taylor & Francis, London and New York, 2001.

Editors Note: This Paper was awarded an "Honorable Mention" in the 2010 Artsutanov Prize competition.

SPACE ELEVATOR: PHYSICAL PROPERTIES AND TRANSPORTATION SCENARIOS

Yashovardhan Sushil Chati,

Department of Aerospace Engineering, Indian Institute of Technology Bombay, Mumbai, India,

yschati@iitb.ac.in

*Georg Herdrich, herdrich@irs.uni-stuttgart.de

*Dejan Petkow, petkow@irs.uni-stuttgart.de

*Stefanos Fasoulas, fasoulas@irs.uni-stuttgart.de

*Hans Peter Röser,

roeser@irs.uni-stuttgart.de

*Institut für Raumfahrtsysteme, Universität Stuttgart, Stuttgart, Germany

Abstract: The first part of the paper deals with the analysis of the physical properties of the space elevator, balanced about the stationary/Lagrangian radius. These properties include tensile stress in a free standing elevator, cross section area of a tapered elevator, cross section area required at the surface, height of the elevator, mass and volume of the material required to construct the elevator, counterweight mass, climber velocity permissible, and energetic cost of launching from the elevator. Advanced materials, having high specific strength, which currently exist (like steel, graphite, Kevlar, Vectran, Zylon, M5, UHMWPE/Dyneema) are considered. Elevators on the Earth, Mars and the Moon are studied. It is concluded that construction of an Earth elevator with the considered materials is infeasible. A Martian elevator appears to be feasible with materials like Zylon. A Moon elevator is possible with all the materials stated above (except steel), particularly Kevlar, Vectran, Zylon, M5, and Dyneema. All the analysis is taking into account a high safety margin of 2. The energy cost analysis indicates that the cost of launching a 1 kg payload to the stationary/Lagrangian orbit is very less in terms of the launch costs involved in rocket propulsion. The second part of the paper studies some mission scenarios involving the space elevator and their trajectory analysis. The scenarios are studied, mainly from a point of view of reducing the fuel requirements (Δv). Time is not of a primary concern as the payload is assumed to be unmanned. Some scenarios studied are using the elevator to establish an orbit of the desired altitude and inclination about the parent body (e.g. Earth), about another body (lunar elevator used to establish Earth orbit, Mars mission), to have rendezvous about elevators on two different bodies so that re-entry to the surface is possible (without the problem of heating). Another innovative scenario is using the elevator to dispose of hazardous material into the Sun. It is found that an elevator a few hundred thousand km above the surface is sufficient for most scenarios. Within this work, only equatorial space elevators are considered.

Numerical values [1], [2], [3], [4]

1. Universal Gravitational constant, $G = 6.674 \times 10^{-11}$ Nm^2/kg^2
2. Gravitational parameter of the Sun, $\mu_{Sun} = 1.33 \times 10^{20}$ Nm^2/kg
3. Gravitational parameter of the Earth, $\mu_{Earth} = 4 \times 10^{14}$ Nm^2/kg
4. Radius of the Earth, $R_{Earth} = 6371000$ m
5. Rotational angular velocity of the Earth, $\omega_{Earth} = 7.29 \times 10^{-5}$ rad/s
6. Axial obliquity of the Earth = $23.44°$
7. Semi major axis of the Earth from the Sun, $a_{Earth} = 1.5 \times 10^{11}$ m

8. Radius of sphere of influence of the Earth, $R_{SOI,Earth}$ = 955000 km
9. Gravitational parameter of Mars, μ_{Mars} = 4.28x10^{13} Nm^2/kg
10. Radius of Mars, R_{Mars} = 3397000 m
11. Rotational angular velocity of Mars, ω_{Mars} = 7.09x10^{-5} rad/s
12. Axial obliquity of Mars = 25.19°
13. Semi major axis of Mars from the Sun, a_{Mars} = 2.28x10^{11} m
14. Inclination of the Martian orbital plane with the ecliptic plane = 1.85°
15. Gravitational parameter of the Moon, μ_{Moon} = 4.91x10^{12} Nm^2/kg
16. Radius of the Moon, R_{Moon} = 1737400 m
17. Rotational angular velocity of the Earth – Moon system about centre of mass, ω_{Moon} = 2.67x10^{-6} rad/s
18. Semi major axis of the Moon from the Earth, a_{Moon} = 384410000 m
19. Inclination of lunar orbital plane with the ecliptic plane = 18.3° (minimum)

A parameter without the subscript of the name of a heavenly body represents the general parameter, unless otherwise explicitly specified. For example, 'R' without any subscript represents the radius of a heavenly body in general, without any specific reference to Earth / Mars / Moon / Sun. Similar connotation applies to other parameters like 'a', 'ω', 'g', etc.

1.   Elevator physical properties

The planetary elevator is constructed balanced about the stationary radius [13,14]. The moon elevator is constructed balanced about the L_1/L_2 Lagrangian points of the moon planet system [15]. The advanced materials in table 1 are considered for construction.

Table 1: Material properties [5] to [12]

Material	Tensile strength, GPa	Density, kg/m³	Young's Modulus, GPa
Steel (Piano wire) [5], [6]	2.3	7900	200
Graphite (AS4) [7]	4.3	1800	10
High Grade Aramid (Kevlar) [8]	3.2	1450	120
Vectran [9]	2	1400	75
Zylon [10]	5.8	1560	280
Magellan3d polymer, M5 [11]	4	1700	270
Dyneema [12]	2.5	975	110

A constant cross section area elevator is not possible due to the very high tensile stresses developed, more than the tensile strength of the above materials.   Hence, the elevator is given a tapered cross section which increases exponentially from the surface to a maximum at the stationary radius/Lagrangian points and then decreases till the top.   This keeps the tower in a state of constant tensile stress, which is controllable. For a tower with the same cross section area at the top and the surface, the height comes out to be 144190 km for the terrestrial elevator, 65786 km for the Martian elevator, 290391 km for the lunar L_1 and 677293 km for the lunar L_2 elevator. The height can be shortened by a counterweight. For most transportation scenarios from the terrestrial and lunar L_1 elevators, heights of 100000 km and 250000 km, respectively, are sufficient. No counterweight is needed for the Martian elevator. To terminate the L_2 elevator within the Earth's sphere of influence, a height of 500000 km is chosen. It is found that the counterweight mass tends to infinity towards the stationary/Lagrangian height and exponentially falls to 0 kg with increasing height. The material mass required increases with height. The total mass is dominated by the counterweight mass and hence, has a trend similar to that shown by the counterweight mass. The material volume is the mass divided by the density. The cross section at the surface should be such that the tension there is able to lift a climber of the desired mass. While ascending or descending the elevator, the elevator tilts due to the Coriolis Effect. It may be a desire to put the maximum deflection (θ) under limit. This puts a constraint on the maximum climber velocity possible, by the following inequality

$$V_{climber} < \frac{g - \omega^2 R}{\omega} \sin\theta \ . \tag{1}$$

The above characteristics were worked out for the different materials, keeping the tensile stress at 50% of the material tensile strength. None of the above materials are suitable for the construction of a terrestrial elevator due to the very high taper ratios, masses and material volume required. For Mars, Zylon is a suitable material. It gives a taper ratio of 165, a material mass of about 890 tonnes and a volume of about 570 m^3. For the lunar L_1/L_2 elevators, except steel, all the above mentioned materials give reasonable values. The taper ratios are less than 50, the conterweight mass is less than 50 tonnes for the L_1 and less than 150 tonnes for the L_2 elevator. The material mass and volume is a few hundred tonnes and m^3, respectively, except for Vectran for which it is a few thousand tonnes, m^3. All the surface cross section areas are of the order of 0.1 mm^2, which is very small. The maximum climber velocities, for a permissible deflection of 1^0, are as high as 2.3 km/s for the terrestrial, 910 m/s for the Martian and 10.6 km/s for the lunar elevators. These high values indicate that the limit on climber velocities will be put by the power systems, not the elevator mechanics.

The main advantage of a space elevator is the very less energy requirement for launch. Unlike a conventional rocket, energy is required only to carry the payload to the launch height. The velocity due to the rotation of the planet/moon is used for injection. So the net energy required for launch is just the difference between the potential energies at the launch height and the surface. The energy requirements for launch of a 1 kg payload from the stationary/Lagrangian height are just 14.8 kWh for the terrestrial, 2.92 kWh for the Martian, 0.7 kWh for the lunar L_1 and 0.8 kWh for the lunar L_2 elevators. The maximum velocity possible from the terrestrial elevator is 7.76 km/s (with respect to the Earth, from a height of 100000 km) and 4.9 km/s from the Martian elevator (with respect to Mars, from a height of 65786 km). These velocities are more than the escape velocities from that height. The maximum velocities from the top of the L_1 (height 250000 km) and L_2 (height 500000 km) elevators are 342 m/s and 2.35 km/s, respectively, with respect to the Earth – Moon system centre of mass.

2. Orbits possible from the planetary space elevator

At any point along the height of the planetary elevator, there is a tangential velocity due to the rotation of the planet. This velocity can be used to inject any payload into an orbit about the planet on which the elevator is built, without requiring any additional thrust. Since the injection velocity is perpendicular to the radius vector from the planet, the injection point is either the periapsis or the apoapsis. Also, the injection velocity is related to the release radius (radius at which it is desired to inject the payload) via the planet's angular rotation velocity. Hence, unlike conventional rocket injection, in which the injection velocity and the injection radius are two independent parameters, for a space elevator injection, there is only one parameter – the release radius (r).

Below the stationary radius, the elevator velocity is less than the circular orbital velocity, leading to the creation of an elliptical orbit. The injection point is the apoapsis of the elliptical orbit. The elevator is in a circular orbit in the stationary radius. Above the stationary radius, the elevator velocity is more than the circular orbital velocity. Hence, above the stationary radius and below the escape radius (where the tower velocity is $\sqrt{2}$ times the orbital velocity), again an elliptical orbit is formed but this time with the injection point as the periapsis. At the escape radius, a parabolic trajectory is formed. Above the escape radius, a hyperbolic trajectory arises. The semi major axis (a) and eccentricity (e) are functions of the release radius (r) and given by

$$a = \left| \frac{-0.5\mu}{\frac{\omega^2 r^2}{2} - \frac{\mu}{r}} \right|, \tag{2}$$

$$e = \left| \frac{\omega^2 r^3}{\mu} - 1 \right|. \tag{3}$$

3. Orbits possible from the lunar L_1 elevator

When the release point is within the Moon's sphere of influence, an elliptical orbit is formed about the Moon. The release point forms the apolune. The orbit lies wholly within the Moon's sphere of influence. Hence, it is stable with respect to perturbations due to the Earth's gravity. When the release point is outside the Moon's sphere of influence, an elliptical orbit is formed about the Earth. The release point forms the apogee. The orbit lies wholly within the Earth's sphere of influence and outside the Moon's sphere of influence. Hence, it is stable with respect to perturbations due to the Moon's gravity. The L_1 elevator cannot provide escape trajectories from either the Moon or the Earth. Also, there is no height from which a circular orbit is possible.

4. Orbits possible from the lunar L_2 elevator

When the release point is within the Moon's sphere of influence, an elliptical orbit is formed about the Moon. The release point forms the apolune. The orbit lies wholly within the Moon's sphere of influence. When the release point is outside the Moon's sphere of influence, till a height of about 96210 km above the lunar surface, an elliptical orbit is formed about the Earth. The release point forms the perigee. The orbit lies wholly outside the Moon's sphere of influence. However, all the orbits extend out beyond the Earth's sphere of influence into the Sun's sphere of influence. Hence, the orbits are unstable with respect to perturbations due to the Sun's gravity and care must be taken regarding these orbits. Beyond 96210 km, hyperbolic trajectories around the Earth are formed.

These trajectories lie completely outside the Moon's sphere of influence. An important thing to be noted is that the sphere of influence concept is just an approximation, strongly valid very far inside the sphere. Actually, the orbit is determined by both the Earth's and the Moon's gravity (and the Sun's if applicable), especially close to the spheres of influence. The sphere is just a mathematical demarcation for ease of analysis.

5.   Planetary elevator to establish orbits about the parent planet

The Δv and Minimum Time To Travel (MTTT) required to establish a circular orbit of the desired altitude and inclination about the parent planet, are studied. The results are studied for only the terrestrial elevator. Two scenarios are analysed:

a) The payload is given a velocity impulse thereby directly injecting it into circular orbit at the desired altitude.

b) The Hohmann type transfer is used.

5.1   Payload injection into circular orbit by velocity impulse

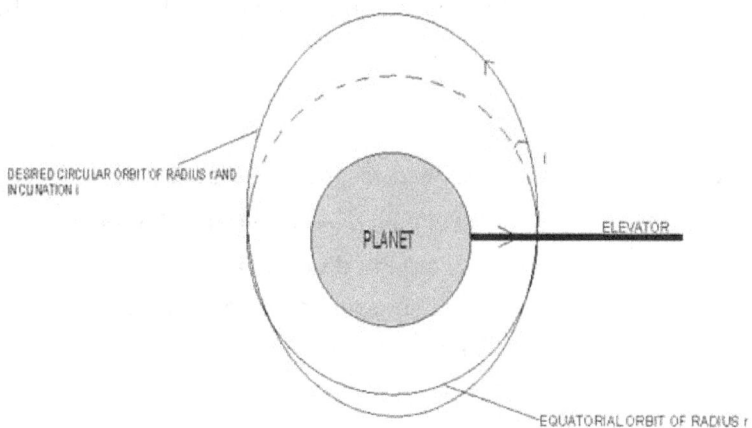

Fig.1 Approximate scheme for single impulse direct injection manoeuvre into the desired circular orbit

The elevator is in orbit at the stationary height. Below the stationary height, the elevator has sub orbital velocity. Therefore, to launch a payload into circular orbit, the velocity has to be increased (prograde firing of rockets). Above the stationary height, the elevator has super orbital velocity. Therefore, for injection, the velocity has to be decreased (retrograde firing of rockets). The velocity imparted by the elevator at radius r is given by

(4)

$$v = \omega r.$$

The orbital velocity at radius r is given by

$$v = \sqrt{\frac{\mu}{r}}.$$ (5)

To achieve a circular orbit by direct injection therefore, implies giving a Δv equal to the difference of the above two equations.

The equatorial elevator can launch payloads into equatorial orbits only. So a further velocity impulse needs to be given to change the orbital plane, to achieve the desired inclination. The entire manoeuvre is achievable by a single impulse to inject the payload directly into the circular orbit of the desired radius r and inclination i (henceforth referred to as single impulse direct injection manoeuvre). A single impulse in the appropriate direction is given to impart the circular shape and inclination simultaneously. The Δv is given by the magnitude of the vectorial difference of the final velocity vector (with circular velocity given by eq. 5) and initial velocity vector (with velocity given by eq. 4). Both the vectors are inclined at an angle i to each other.

$$\text{Total } \Delta v = \sqrt{(\frac{\mu}{r} + \omega^2 r^2 - 2\sqrt{\frac{\mu}{r}} \times \omega r \times \cos(i))} \tag{6}$$

MTTT is the time required to carry the payload up to the desired height along the elevator.

$$\text{MTTT} = \frac{r-R}{v_{asc}} \tag{7}$$

where v_{asc} is the velocity at which the payload ascends the elevator.

As the desired radius increases, the MTTT increases linearly. There is no dependence on the inclination because change of orbital plane is assumed to be an impulsive maneuver. The MTTT for a typical Low Earth Orbit (LEO) (altitude of 1000 km say) at a climber ascent velocity of 70 m/s is just 4 hours and that for a 100000 km altitude orbit is 16.5 days.

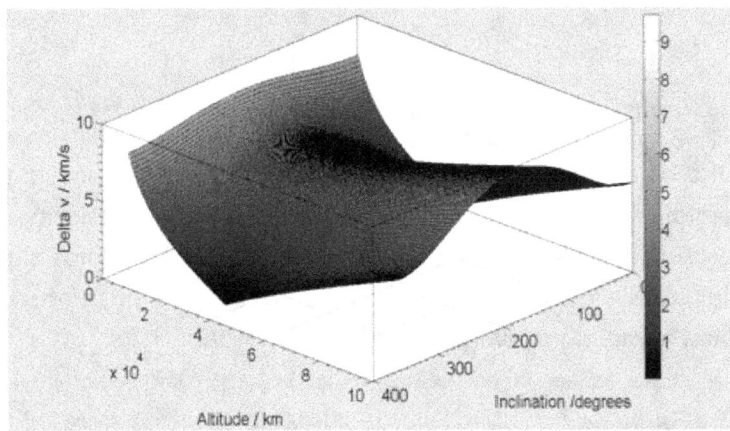

Fig.2 Terrestrial elevator: Δv for single impulse direct injection manoeuvre

In Fig.2, the plot is symmetric about i of 180°. This is expected due to the periodic nature of the inclination. There is a local maximum at this angle. There is also a local minimum at the stationary altitude (35850 km). This is expected as at the stationary altitude, the elevator is already in a circular orbit. So the Δv needed is just to change the orbital plane. At i of 0° and altitude equal to stationary altitude, the Δv is actually equal to 0 m/s.

5.2   Hohmann type transfer

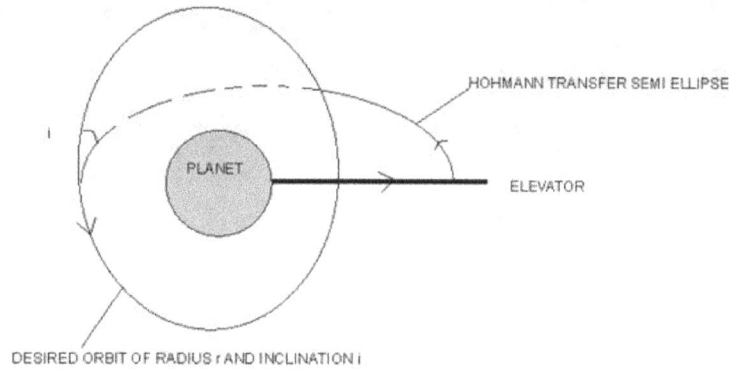

Fig.3 Approximate scheme for the Hohmann type transfer to establish the desired circular orbit

First the radius along the elevator at which the payload, if released, travels along an elliptical orbit, whose other apside distance is equal to the radius of the desired orbit r, is calculated. This radius is called the release radius, r_{rel}. It is given by the real, positive solution of the following equation:

$$\omega^2 r_{rel}^4 + \omega^2 r \times r_{rel}^3 - 2\mu r = 0 \tag{8}$$

There is no dependence of r_{rel} on the orbit inclination.

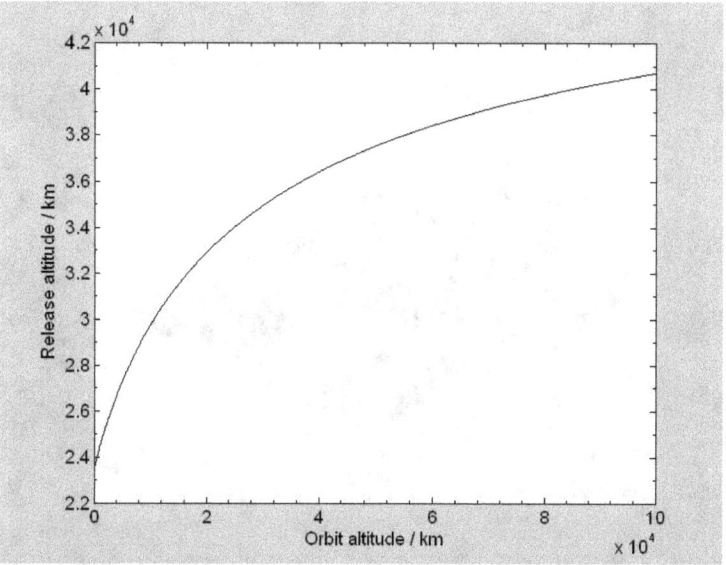

Fig.4 Hohmann type transfer: Release altitude from terrestrial elevator vs orbit altitude

In Fig.4, it is seen that the release altitude goes on increasing with the altitude of the orbit. Also, since the elevator is already in circular orbit at the stationary radius, at the stationary radius the release radius is equal to the stationary radius, as expected.

On reaching the other apside, the payload is given a single impulse to change the orbit into a circular orbit of the desired radius r and inclination i. The Δv is given by the magnitude of the vectorial difference of the final velocity vector (circular orbit velocity) and initial velocity vector (with velocity given by the elliptical orbit velocity at the apside). Both the vectors are inclined at an angle i to each other.

Elliptical orbit velocity at radius r,

$$v_1 = \sqrt{2\mu(\frac{1}{r} - \frac{1}{r+r_{rel}})}. \tag{9}$$

$$v_2 = \sqrt{\frac{\mu}{r}}$$

$$\Delta v = \sqrt{v_1^2 + v_2^2 - 2 \times v_1 \times v_2 \times \cos(i)} \tag{10}$$

MTTT required is the time taken to carry the payload up to the release radius along the elevator plus the time taken to travel to the other apside of the Hohmann type transfer ellipse

$$\text{MTTT} = \frac{r_{rel}-R}{v_{asc}} + \frac{\pi}{\sqrt{\mu}}(\frac{r+r_{rel}}{2})^{1.5} \tag{11}$$

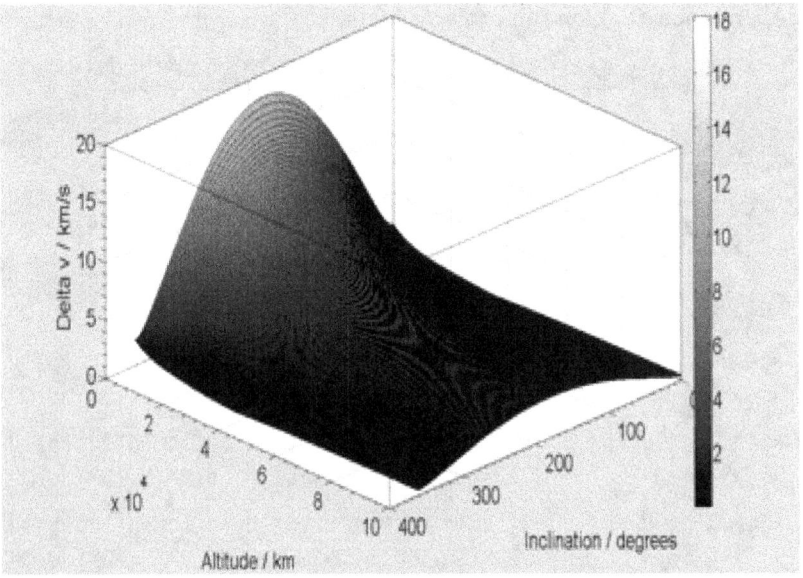

Fig.5 Δv requirement for Hohmann type transfer transfer from the terrestrial elevator

In Fig.5, the plot is again symmetric about i of 180°. There is a local maximum at this angle, as observed in the earlier plot for direct injection too. There is a local minimum at the stationary altitude. This is again expected as at the stationary altitude, the elevator is already in a circular orbit. For an equatorial stationary orbit, the Δv is 0. Lower altitudes require more Δv than higher altitudes.

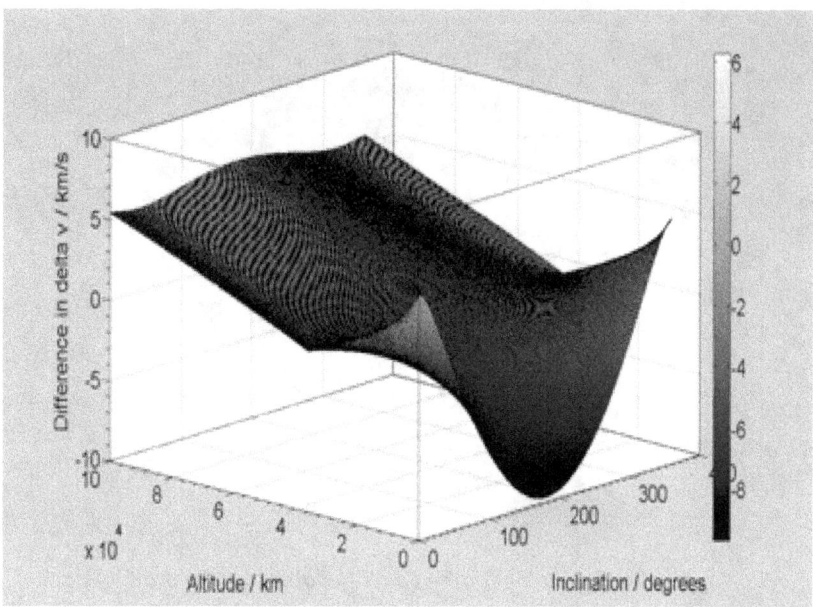

Fig.6 Difference in Δv for direct injection and Hohmann type transfer from the terrestrial elevator

Very interesting things are observed about the plot in Fig.6. Above the stationary radius, the Hohmann type transfer requires less Δv than the direct injection manoeuvre, for all altitudes and inclinations. In fact the Δv difference increases with altitude, making Hohmann type transfer a cheaper manoeuvre for higher altitudes. For any altitude above the stationary height, the difference is maximum at i of 180°. Below the stationary radius the trend changes. The Hohmann type transfer requires less Δv than the direct injection manoeuvre, only at lower inclinations (and corresponding symmetric inclinations). At higher inclinations, the direct injection is a cheaper manoeuvre. The direct injection is the cheapest with respect to the Hohmann type transfer at 180° inclination.

So the Hohmann type transfer is not the minimum Δv manoeuvre always (as is seen in Fig.6 below the stationary height). But according to theory, Hohmann transfer is the minimum Δv manoeuvre. This discrepancy can be resolved by understanding that only the transfers for altitudes above the stationary height are the Hohmann transfers in the true sense. For the Hohmann transfer, the apoapsis of the transfer ellipse should be at the desired radius. This happens only above the stationary height. Below the stationary height, it is the periapsis which is at the desired radius.

The MTTTs for both the transfers monotonically increase with altitude. For altitudes less than about 40000 km (which includes LEOs), the Hohmann type transfer requires more time than direct injection. But for higher altitudes (more than about 40000 km), the Hohmann type transfer not only saves on energy but also on time, as compared to the direct injection manoeuvre.

6.   Planetary elevator to establish missions to other planets

An elevator on a planet can be used to establish orbits around planets other than the parent planet. For example the launch of a satellite from the terrestrial elevator into an orbit around Mars for observational purposes could be a scenario of relevance. Hence, the scenario for transporting a payload up the terrestrial elevator and launching it into a circular orbit of the desired altitude r and inclination i around Mars is analysed. The Hohmann trajectory [17] and the method of patched conics are used for the same.

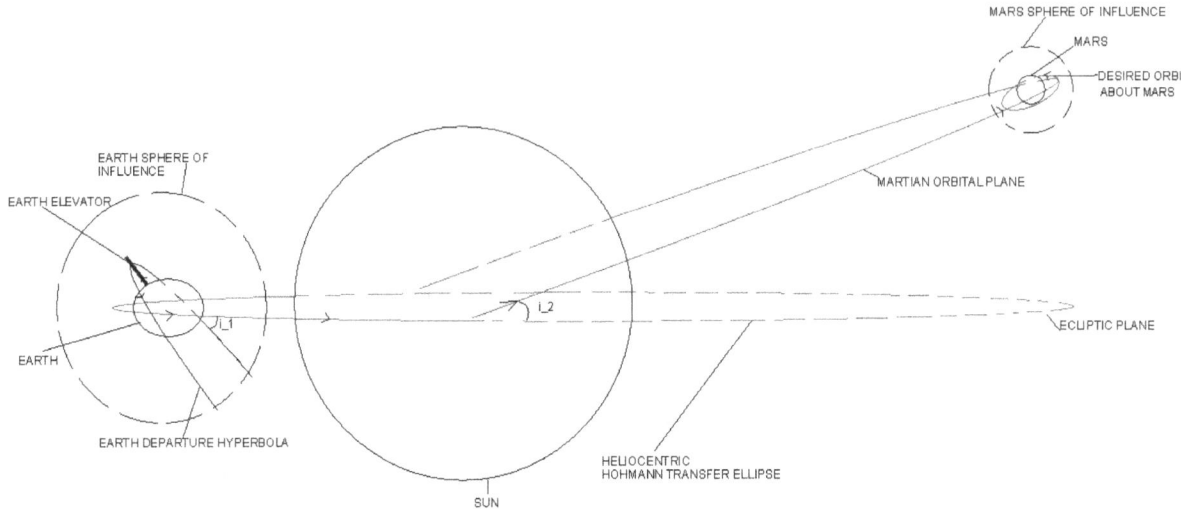

Fig.7 Approximate scheme for the Earth elevator to Mars orbit scenario

Step 1: The payload is injected into a departure hyperbola around the Earth. This hyperbola is also a Hohmann ellipse around the Sun, with periapsis at the Earth and apoapsis at Mars. The important parameter is the hyperbolic excess velocity which is calculated as follows:

Semi major axis of the ellipse around the Sun

$$a = \frac{a_{Earth} + a_{Mars}}{2}. \tag{12}$$

Velocity with respect to the Sun at periapsis

$$v = \sqrt{\mu_{Sun} \left(\frac{2}{a_{Earth}} - \frac{1}{a} \right)}. \tag{13}$$

Hyperbolic excess velocity with respect to the Earth, $v_{\infty, Earth} =$

$$\left| \sqrt{\mu_{Sun} \left(\frac{2}{a_{Earth}} - \frac{1}{a} \right)} - \sqrt{\frac{\mu_{Sun}}{a_{Earth}}} \right|. \tag{14}$$

(Here a circular orbit of the Earth around the Sun is assumed.)

Step 2: The release height from the terrestrial elevator to achieve the hyperbolic excess velocity is calculated. This height is more than the escape height, as the payload is on a hyperbolic path around the Earth. The release radius is the real, positive solution of the following equation:

$$\omega_{Earth}^2 r_{rel}^3 - v_{\infty, Earth}^2 r_{rel} - 2\mu_{Earth} = 0 \tag{15}$$

Plugging the values, the release height from the terrestrial elevator comes out to be 56860 km.

The time required to carry the payload up to the release height along the terrestrial elevator is given by

$$T_1 = \frac{r_{rel} - R_{Earth}}{v_{asc}}.$$ (16)

A thing to be noted is that the payload should be injected into the departure hyperbola at a point of time such that the hyperbola, when brought into the ecliptic plane, has its symmetry axis pointing towards the Sun. The position of Mars during injection should be such that the payload and Mars arrive at the apoapsis of the Hohmann transfer orbit around the Sun at the same time. These conditions determine the launch window. This concept is not dealt with here.

Step 3: The departure hyperbola is in the Earth's equatorial plane. It is first brought into the ecliptic plane by a normal velocity impulse. This is the first Δv in the whole mission.

$$\Delta v_1 = 2v\sin\left(\frac{\theta}{2}\right).$$ (17)

Here, v is the velocity at the point of intersection of the new hyperbola (in the ecliptic plane) and the old hyperbola (in the Earth's equatorial plane).

$$v = \sqrt{\left(v_{\infty,Earth}^2 + \frac{2\mu_{Earth}}{p}\right)}$$ (18)

The parameter p is the semi latus rectum of the hyperbola.

$$p = \frac{\omega_{Earth}^2 r_{rel}^4}{\mu_{Earth}}$$ (19)

The parameter θ is the Earth's axial obliquity.

Next, the Hohmann transfer orbit around the Sun is given a normal impulse to bring it into the Martian orbital plane.

$$\Delta v_2 = 2v\sin\left(\frac{\theta}{2}\right)$$ (20)

Here, v is the velocity at the point of intersection of the new transfer ellipse (in the Martian orbital plane) and the old transfer ellipse (in the ecliptic plane).

$$v = \sqrt{\mu_{sun}\left(\frac{2}{p} - \frac{1}{a}\right)} \tag{21}$$

p is the semi latus rectum of the transfer ellipse.

$$p = \frac{\mu_{Sun}\left(\frac{2}{a_{Earth}} - \frac{1}{a}\right) \times a_{Earth}^2}{\mu_{Sun}} \tag{22}$$

The parameter θ is the angle of Mars orbital plane with the ecliptic plane.

Step 4: The payload, travelling along the Hohmann transfer ellipse, once inside the Martian sphere of influence is on an approach hyperbola around Mars. The hyperbolic excess velocity with respect to Mars is again calculated as follows:

Velocity with respect to the Sun at apoapsis,

$$v = \sqrt{\mu_{sun}\left(\frac{2}{a_{Mars}} - \frac{1}{a}\right)}. \tag{23}$$

Hyperbolic excess velocity with respect to Mars,

$$v_{\infty,Mars} = \left| \sqrt{\mu_{sun}\left(\frac{2}{a_{Mars}} - \frac{1}{a}\right)} - \sqrt{\frac{\mu_{sun}}{a_{Mars}}} \right|. \tag{24}$$

(Here a circular orbit of Mars around the Sun is assumed to simplify matters.)

Step 5: The payload, travelling along the Hohmann transfer ellipse, reaches the apoapsis, near Mars. The time spent in travelling along the transfer ellipse is given by

$$T_2 = \frac{\pi}{\sqrt{\mu_{sun}}} \times (a)^{1.5}. \tag{25}$$

A Δv is now needed to capture the payload into the desired circular orbit about Mars. This Δv is equal to the difference between the velocity at the Martian approach hyperbola periapsis and the desired circular orbit velocity around Mars.

$$\Delta v_3 = \sqrt{v_{\infty,Mars}^2 + \frac{2\mu_{Mars}}{r}} - \sqrt{\frac{\mu_{Mars}}{r}} \tag{26}$$

The final velocity impulse is needed to change the orbital plane of the circular orbit. The Δv required is

$$\Delta v_4 = 2\sqrt{\frac{\mu_{Mars}}{r}}\sin\left(\frac{i - axial\ obliquity\ of\ Mars}{2}\right). \tag{27}$$

$$\text{Total } \Delta v = |\Delta v_1| + |\Delta v_2| + |\Delta v_3| + |\Delta v_4| \tag{28}$$

An additional time is spent in traversing a quarter of the circular orbit in the Martian orbital plane before the plane can be changed to give the final orbit. This time is given by

$$T_3 = \frac{\pi r / 2}{\sqrt{\mu_{Mars}/r}}. \tag{29}$$

$$MTTT = T_1 + T_2 + T_3 \tag{30}$$

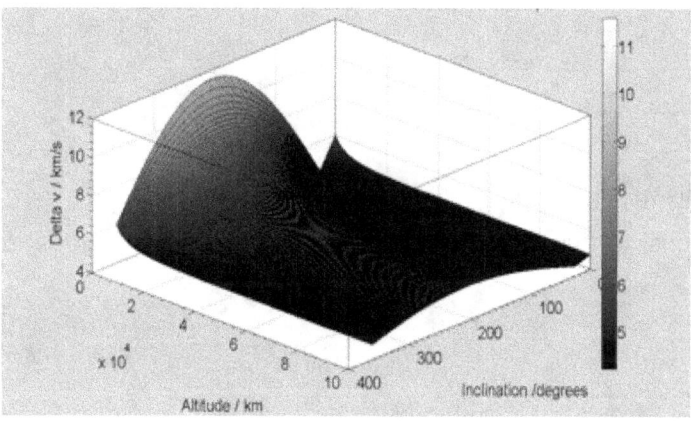

Fig.8 Earth elevator to Mars orbit: Δv required for desired orbit capture

In Fig.8, for any altitude, there is a local maximum at inclination of about 210°. The Δv at this inclination monotonically decreases with altitude, the maximum being as high as about 11.6 km/s (which is also the global maximum). For any altitude, there is a local minimum at inclination about 25°. This is because of the absence of Δv needed to change the inclination (with respect to the Martian equatorial plane) of the final orbit as the axial obliquity of Mars is 25.19°. At this inclination, the Δv shows significantly different behaviour. There is a local minimum at about 9000 km altitude, where the Δv is as low as 4.19 km/s (which is also the global minimum). To see the behaviour with altitude at other inclinations, individual 1d plots need to be studied at these inclinations.

The MTTT monotonically increases with the orbit altitude. An orbit of 1000 km altitude requires some 268 days to complete the mission.

Elevator systems can also be used to transport a payload from the surface of one planet to the other. An important advantage of this is that not only are the Δv requirements less than conventional rocket missions but the problem of atmospheric entry heating is also done away with. The payload simply descends down the elevator to the planet's surface. Here, the Δv and MTTT required for transportation from the Earth to Mars are determined.

Step 1: The payload is first put into a circular, equatorial orbit around Mars so that it can do rendezvous with the Martian elevator. Position rendezvous is assumed to be done as soon as the payload comes into the equatorial orbit (this is just done to simplify calculations. Otherwise the time required would simply increase). A Δv is then given to the payload to achieve velocity rendezvous with the Martian elevator. If the payload does rendezvous with the elevator at radius r, then the Δv required for velocity rendezvous is

$$\Delta v = \omega_{Mars}r - \sqrt{\frac{\mu_{Mars}}{r}}. \tag{31}$$

No additional time is required.

Step 2: The payload descends down the elevator at a constant speed. If this speed is v_{des}, then the time required for descent is

$$T = \frac{r - R_{Mars}}{v_{des}}. \tag{32}$$

No Δv is required in this step.

The total Δv is equal to the $|\Delta v|$ required to achieve circular, equatorial orbit of radius r about Mars

(eq. 28) + the $|\Delta v|$ from eq. 31. (33)

The total MTTT is equal to the MTTT required to achieve circular, equatorial orbit of radius r about Mars

(eq. 30) + T from eq. 32. (34)

In Fig.9, the variation of the total Δv with the Martian elevator rendezvous radius r is plotted to see if there exists a radius at which the total Δv is minimum.

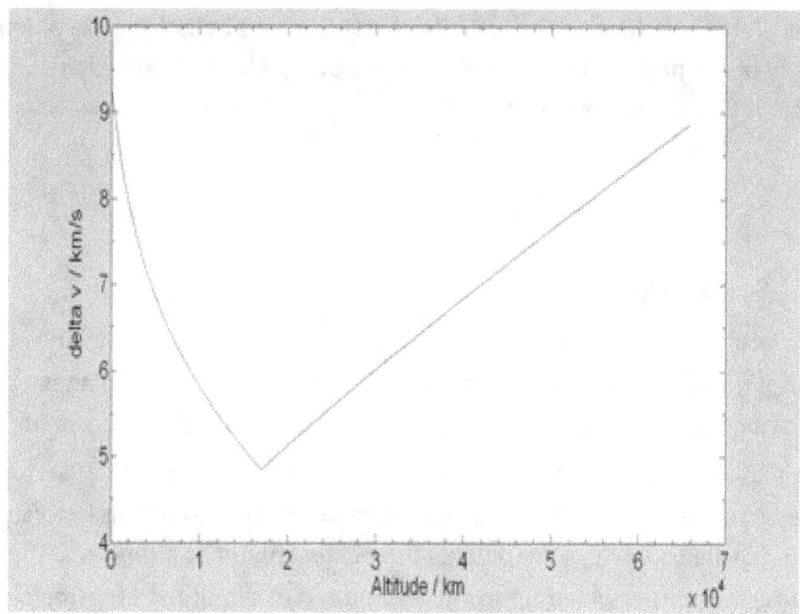

Fig.9 Earth surface to Mars surface: Δv vs Martian elevator rendezvous altitude

In Fig.9, it is seen that the total Δv is minimum when the rendezvous with the Martian elevator takes place at the Martian stationary altitude (17034 km). This is a very important characteristic of elevator to elevator missions. Plugging in the values and using the Martian stationary radius as the rendezvous radius, the minimum Δv for the whole scenario turns out equal to 4.85 km/s and the MTTT as 271.07 days (assuming climber ascent and descent velocities as 70 m/s).

7.   Elevator to drop hazardous material into the sun

This is a special, innovative application, required when it is desired to dispose of hazardous material (like nuclear waste) permanently by dumping it into the Sun where it just burns off completely [14]. For this, the hazardous payload is made to escape the home planet into an orbit around the Sun. If the velocity of the payload is now made to 0, it will simply fall off into the Sun.

The escape radius from a planetary equatorial elevator is given by

$$r_{esc} = \sqrt[3]{\left(\frac{2\mu}{\omega^2}\right)}. \tag{35}$$

The escape velocity at this height from the elevator is hence,

$$v_{esc} = r_{esc} \times \omega = \sqrt[3]{2\mu\omega}. \tag{36}$$

Once released from this height, the payload follows a parabolic trajectory with respect to the planet. The velocity of the payload with respect to the planet, when it reaches the edge of the planet's sphere of influence, is v which is given by

$$v = \sqrt{\frac{2\mu}{R_{SOI,Planet}}} \qquad (37)$$

where $R_{SOI, Planet}$ is the radius of the sphere of influence of the planet.

The velocity with respect to the Sun (v_{net}) is maximum when the velocity of the planet with respect to the sun (v_s) and the velocity of the payload with respect to the planet (v) are in the same direction. The net velocity is minimum when the two velocities are in the opposite direction.

This velocity is too large to be achieved in an impulse. Since time is not of serious concern, a low thrust manoeuvre can be used. The thrusters are assumed to give a constant, low value of acceleration A in the direction opposite to travel, once outside the planet sphere of influence. The MTTT is given by the time required to carry the payload up to the escape radius along the elevator plus the time taken to reach the edge of the sphere of influence along the parabolic orbit plus the time taken to reduce the velocity to zero in low thrust manoeuvre.

$$MTTT = \frac{r_{esc} - R}{v_{asc}} + \frac{1}{3}\sqrt{\frac{p^3}{\mu}} \frac{(2+\cos\theta)\sin\theta}{(1+\cos\theta)^2} + \frac{v_{net}}{A} \qquad (38)$$

The parameter p is the semi latus rectum of the parabolic orbit and θ is the polar angle at the edge of the sphere of influence.

For an Earth based elevator, solving the above equations gives an MTTT of 50 to 52 days if the thrusters give a deceleration of 0.01 m/s². The payload follows a "spiral" path and drops into the Sun.

8.   Lunar elevator to establish missions to the earth

The scenario of transporting a payload up the lunar L_1 elevator to put it in an orbit around the Earth is analysed [16]. The cargo can be lunar regolith which is rich in minerals. It is desired to establish a circular orbit of radius r and inclination i around the Earth. This can be done in the following way:

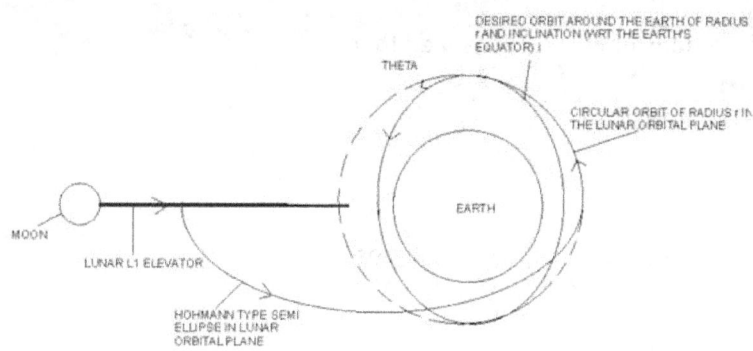

THETA : ANGLE BETWEEN DESIRED EARTH
ORBIT AND LUNAR ORBITAL PLANE

DESIRED ORBIT AROUND THE EARTH OF RADIUS
r AND INCLINATION (WRT THE EARTH'S
EQUATOR) i

THETA

CIRCULAR ORBIT OF RADIUS r IN
THE LUNAR ORBITAL PLANE

EARTH

MOON

LUNAR L1 ELEVATOR

HOHMANN TYPE SEMI
ELLIPSE IN LUNAR
ORBITAL PLANE

Fig.10 Approximate scheme for the lunar L₁ elevator to the Earth circular orbit

Step 1: First, the release radius along the L_1 elevator is calculated so that the perigee of the resulting elliptical orbit lies at the desired orbit radius r. This release point necessarily lies beyond L_1. The release radius, r_{rel} (from the Moon's centre) is given by the real, positive root of the following equation:

$$\omega_{Moon}^2 t^4 + \omega_{Moon}^2 rt^3 - 2\mu_{Earth}r = 0, \; r_{rel} = a_{Moon} - t \qquad (39)$$

The release radius depends only on the orbit radius and not on its inclination. The release height is observed to monotonically decrease with the orbit altitude.

The time elapsed till now is

$$T_1 = \frac{r_{rel} - R_{Moon}}{v_{asc}}. \qquad (40)$$

Step 2: A Δv is required to be given at the transfer ellipse perigee to change the orbit into a circular orbit of radius r.

$$\Delta v_1 = \sqrt{\frac{\mu_{Earth}}{r}} - \sqrt{2\mu_{Earth}\left(\frac{1}{r} - \frac{1}{t+r}\right)} \qquad (41)$$

No additional time is required.

Step 3: The orbit plane is now changed from lunar orbital plane to the desired inclination (with respect to the Earth's equator). This is done by either giving a normal impulse to the transfer ellipse itself (before step 2 is carried out) at the point of intersection of the old orbit (in the lunar orbital plane) and the new orbit (at the desired inclination) or by giving a normal impulse to the circular orbit of radius r, lying in the lunar orbital plane (after step 2 is carried out). In either case, the orbital plane has to be changed by an angle θ = Desired inclination, i – angle of lunar orbital plane to Earth's equatorial plane.

If the plane of the transfer ellipse is changed then

$$\Delta v_{21} = 2v\sin(\tfrac{\theta}{2}). \tag{42}$$

Here v is velocity at node of old and new ellipses.

$$v = \sqrt{2\mu_{Earth}(\tfrac{1}{x} - \tfrac{1}{t+r})} \tag{43}$$

Here, x is the distance from the Earth to that node of the new and old elliptical orbits which is nearer to the injection point in the direction of travel. It is given as follows:

Semi latus rectum of the elliptical orbit,

$$p = \frac{\omega_{Moon}^2 t^4}{\mu_{Earth}}. \tag{44}$$

If φ is the angle made by the nodal line vector (pointing towards the ascending node) with the perigee (measured in the anti-clockwise direction), then the angle of the node, nearer to the point of injection in the direction of travel, with the perigee is either ψ = 180° + φ (if φ < 180°) or simply ψ = φ (if φ >= 180°).

$$x = \frac{p}{1+(\tfrac{2t}{t+r} - 1)\times\cos\psi}. \tag{45}$$

No additional time is required as the orbit change takes place during the course of the transfer along the ellipse.

$$T_2 = 0 \tag{46}$$

If the plane of the final circular orbit is changed, then

$$\Delta v_{22} = 2v\sin(\tfrac{\theta}{2}). \tag{47}$$

Here v is the circular orbit velocity around the Earth.

Additional time is spent in traversing a part of the circular orbit till the payload reaches the point of intersection (node) of the old and new circular orbits. This time is given by

$$T_2 = \frac{r \times \psi \times {}^\pi/_{180}}{\sqrt{{}^{\mu_{earth}}/_r}} \, .$$
(48)

Here ψ is angle of the node, nearer to the perigee in the direction of travel, measured anticlockwise from the perigee.

ψ = ϕ, if $\phi < 180°$

      = ϕ - $180°$, otherwise

That manoeuvre is chosen which gives the lesser Δv i.e.

Δv_2 = minimum (Δv_{21}, Δv_{22})
(49)

In the present analysis, $\phi = 90°$.

The total Δv is

$| \Delta v_1| + | \Delta v_2|$.
(50)

MTTT required is

$T_1 + T_2$.
(51)

The Δv variation is very similar to that for the Earth elevator to Mars orbit scenario. For any altitude, there is a local maximum at inclination about 200°. The Δv at this inclination monotonically decreases with altitude, the maximum being very high at 18.5 km/s (which is also the global maximum). For any altitude, there is a local minimum at inclination about 18°. This is because of the absence of Δv needed to change the inclination (with respect to the Earth equatorial plane) of the final orbit, which is at an angle of 18.3° to the lunar orbital plane (this angle actually regresses over time, but the minimum angle is chosen as a conservative estimate). At this inclination, the Δv again monotonically decreases with altitude. At low altitudes, the Δv is about 3.1 km/s which falls down to about 0.5 km/s at an altitude of 100000 km. To see the behaviour with altitude at other inclinations, individual 1d plots need to be studied at these inclinations.

The MTTT monotonically falls with the altitude. Typical LEO scenarios require slightly more than a month for completion.

Elevator systems (terrestrial and lunar L_1 elevators) can be used to transport a payload from the surface of the Moon to that of the Earth. As explained for the Earth – Mars case, the payload is first put into a circular, equatorial orbit around the Earth so that it can do rendezvous with the terrestrial elevator. Δv is then given to the payload to achieve velocity rendezvous with the terrestrial elevator. The payload then descends down the terrestrial elevator at a constant speed. The total Δv is minimum when the rendezvous with the terrestrial elevator takes place at the Earth stationary altitude. Plugging in the values, the minimum Δv for the whole scenario turns out to be equal to 1.89 km/s and the MTTT as 30.7 days (assuming climber ascent and descent velocities as 70 m/s).

9. Lunar elevator to establish missions to Mars

The lunar elevator can be used to send a payload into an orbit around Mars. The mission is similar to the mission for putting a payload into a Martian orbit using the terrestrial elevator. The departure hyperbola (whose parameters are the same as in eq. 12 - 14) lies in the lunar orbital plane. It needs to be brought into the ecliptic plane. This is done at the point of intersection of the old hyperbola (in the lunar orbital plane) and the new hyperbola (in the ecliptic plane). The point of intersection lies at a distance from the Earth, equal to the hyperbola semi latus rectum. For ease of analysis, the condition that the point of intersection lies within the Earth's sphere of influence is enforced. This puts a limit on the payload release radius along the lunar elevator. The governing equation (where r_{rel} is the distance from the Earth to the release point) is

$$v_{\infty,Earth}^2 r_{rel}^2 + 2\mu_{Earth} r_{rel} - \mu_{Earth} R_{SOI,Earth} < 0 \qquad (52)$$

Solving this inequality by putting in the values, the condition is that the release radius should be less than 169054 km from the Earth's centre. This corresponds to a release height greater than 213619 km above the lunar surface along the L_1 lunar elevator. Since the favourable release points all lie on the L_1 elevator from which Earth escape hyperbolae are not possible without additional thrust, a Δv has to be given during injection to put the payload into the appropriate departure hyperbola around the Earth. More the release height from the lunar surface, more is this additional Δv. So the release height is kept as less as possible and satisfying the constraints. A release height of 215000 km above the lunar surface along the L_1 elevator is chosen. This is equal to a distance from the Earth's centre (r_{rel}) of 167673 km.

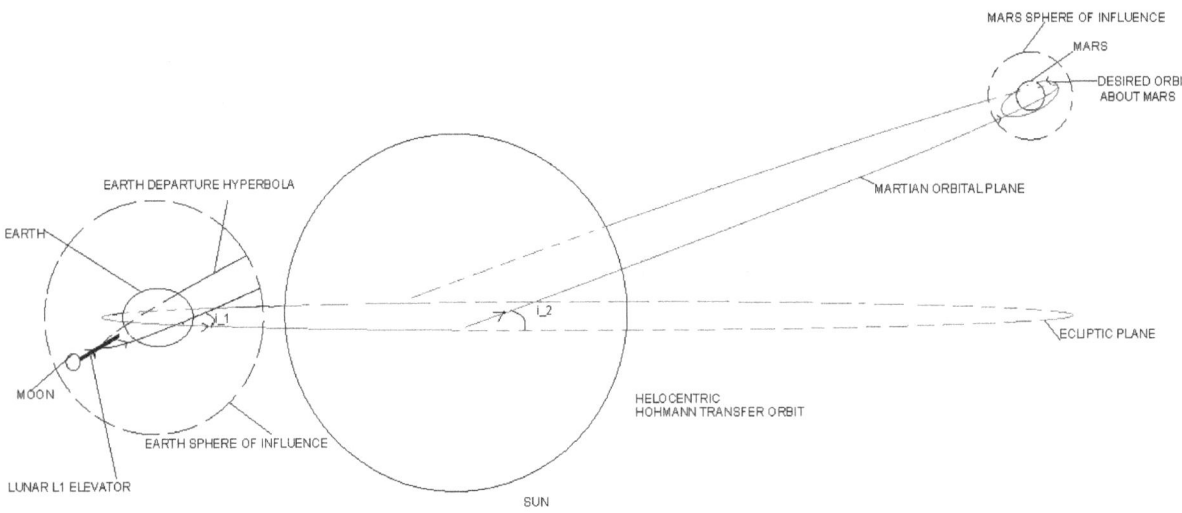

Fig.11 Approximate scheme for lunar L_1 elevator to Mars orbit scenario

Step 1: A Δv during injection, equal to the difference between the perigee velocity of the departure hyperbola and the L_1 elevator velocity at the release height is given.

$$\Delta v_0 = \sqrt{\left(v_{\infty,Earth}^2 + \frac{2\mu_{Earth}}{r_{rel}}\right)} - \omega_{Moon} r_{rel} \tag{53}$$

The time elapsed till now is the time needed to carry the payload up to the release height.

$$T_1 = \frac{a_{Moon} - R_{Moon} - r_{rel}}{v_{asc}} \tag{54}$$

Step 2: The departure hyperbola is in the lunar orbital plane. A normal impulse is given to bring it into the ecliptic plane. This impulse is given at the point of intersection of the old hyperbola (in the lunar orbital plane) and the new hyperbola (in the ecliptic plane). The distance of the point of intersection from the Earth is the semi latus rectum. By the condition enforced through eq. 52, the point of intersection lies within the Earth's sphere of influence and hence, the only influence is of the Earth's gravity. Then the orbit is brought into the Martian orbital plane. The entire manoeuvre and the steps following it are similar to the procedure adopted in sec. 6 (step 3 onwards) and hence, are not explained again.

The total Δv is

$$|\Delta v_0| + |\Delta v| \text{ from step 3 onwards of sec.6.} \tag{55}$$

MTTT is

$$T_1 + \text{MTTT from step 3 onwards of sec. 6.} \tag{56}$$

The variation of the Δv with the desired Martian orbit altitude and inclination is very similar to that for the terrestrial elevator to Mars orbit mission. The global maximum is 13.5 km/s and the global minimum is 6.25 km/s. The MTTT monotonically increases with the orbit altitude.

Payloads can also be transferred between the lunar and the Martian elevators. The same steps as those for the Earth – Mars transfer are followed. The minimum Δv is when the rendezvous takes place at the Martian stationary radius. For this rendezvous radius, plugging in the values, a minimum Δv of 6.91 km/s and corresponding MTTT of 297.22 days are obtained.

10.   Conclusions

An Earth bound elevator seems impossible with the currently available advanced materials. A Martian elevator is also taxing on the materials, though some of the existing materials (like Zylon) might be strong and light enough to give meaning to its construction. A Moon bound tapered elevator, made of existing high strength composites and polymers, with a constant tensile stress of 50% of the material tensile strength, is possible. An energy analysis is done. From the launch point of view, it is seen that the space elevator is a very cheap means of transport, way cheaper than the current rockets. For example, the Apollo 16 required Δv of 17.63 km/s for the Earth – Moon mission [16]. However a terrestrial – lunar L_1 elevator scenario can achieve this in a Δv of only about 2 km/s. This difference brings out the huge energy savings, at least qualitatively. Only manoeuvres involving creation of orbits with a desired radius and inclination are considered. Other orbital elements (like longitude of ascending node, argument of periapsis) can be analysed in a similar way. Elevators can be used for launching payloads into orbits about other bodies. They can also be used to bring payloads down to the surface at slow constant speeds. They thus completely do away with the problem of aerodynamic heating due to the large re - entry speeds. Analysis of release radii for the Earth and the Moon elevators shows that an elevator of length of the order of a few hundred thousand km above the surface is sufficient for most scenarios. The analysis done is only preliminary to get some quantitative engineering estimates for the energy and time requirements. It needs to be further analysed and simulated before any practical mission can be carried out. Transportation scenarios' analysis doesn't take into account any of the material properties of the elevator. It is concerned only with the available height of the elevator. Hence, it is completely decoupled from the analysis of elevators from a materials point of view.

Acknowledgment

The first author would like to thank the German Academic Exchange Service (DAAD) for funding his stay in Stuttgart during the summer of 2010.

References

[1] "Sun", http://en.wikipedia.org/wiki/Sun (last accessed on 26 June 2010)

[2] "Earth", http://en.wikipedia.org/wiki/Earth (last accessed on 26 June 2010)

[3] "Mars", http://en.wikipedia.org/wiki/Mars (last accessed on 26 June 2010)

[4] "Moon", http://en.wikipedia.org/wiki/Moon (last accessed on 26 June 2010)

[5] "Tensile strength", http://en.wikipedia.org/wiki/Tensile_strength (last accessed on 15 June 2010)

[6] "Young's modulus", http://en.wikipedia.org/wiki/Young's_modulus (last accessed on 15 June 2010)

[7] www.cevp.co.uk/general-graphite.htm (last accessed on 15 June 2010)

[8] "Aramid High Modulus Grade", http://www.matbase.com/material/fibres/synthetic/aramid-high-modulus-grade/properties (last accessed on 15 June 2010)

[9] "Tensile Properties", http://www.vectranfiber.com/pdf/8Pages from Vectran_broc_final61206-8.pdf (last accessed on 15 June 2010)

[10] Huang, Y.K., Frings, P.H. and Hennes, E., "Mechanical properties of Zylon/epoxy composite", Composites Part B: engineering, Vol. 33, pp. 109 – 115, 2002

[11] Cunniff, Philip M., Auerbach, Margaret A., Eugene, Vetter, Sikkema Doetze J., "High Performance "M5" Fiber For Ballistics/Structural Composites", 23rd Army Science Conference, 2004

[12] "Dyneema",http://www.matbase.com/material/fibres/synthetic/dyneema/properties (last accessed on 15 June 2010)

[13] Aravind, P.K., "The Physics of the Space Elevator", American Journal of Physics, Vol. 75, No. 2, pp. 125 – 130, Feb 2007

[14] Pearson, Jerome, "The orbital tower: a spacecraft launcher using the Earth's rotational energy", Acta Astronautica, Vol. 2, pp. 785 – 799, 1975

[15] Pearson, Jerome, "Anchored Lunar Satellites for Cislunar Transportation and Communication", The Journal of the Astronautical Sciences, Vol. 27, No. 1, pp. 39 – 62, January – March, 1979

[16] Engel, Kilian A., "Lunar transportation scenarios utilising the Space Elevator", Acta Astronautica, Vol. 57, pp. 277 – 287, 2005

[17] "Which Way to Mars, Trajectory analysis", Kelsey B. Lynn, http://www.mscd.edu/~eas/Goedecke/asci512/Additional%20512%20Websites/512%20extra%20materials/512_day3/Mars%20Trajectories.htm (last accessed on 26 June 2010)

[18] Chati, Y.S., Herdrich, G., Petkow, D., Fasoulas, S., Röser, H.-P., "Space Elevator: Analysis of Physical Properties and Mission Scenarios", Internal Report IRS-10-IB07, Institut für Raumfahrtsysteme, Universität Stuttgart, Stuttgart, June 2010 (available upon request).

HEAT DISSIPATION ISSUES IN SPACE ELEVATOR CLIMBERS

Gaylen Hinton
gaylenhinton@yahoo.com

Gaylen Hinton
gaylenhinton@yahoo.com

Gaylen Hinton
gaylenhinton@yahoo.com

Abstract: There is going to be a tremendous amount of energy involved in the ascent of a laser powered climber on a space elevator (SE) tether. A significant portion of that energy will result in a heat load on the climber that must be dissipated. The only way that heat can be dissipated in space is through black-body radiation. That typically involves a radiator surface that can radiate out more energy than it receives through solar radiation at its operating temperature. In order to dissipate large amounts of heat energy in space, either very large surface areas or very high operating temperatures are needed. However, with photovoltaics, the operating temperature would not likely exceed 100° C. As a result, the heat dissipation requirements of a laser powered SE climber will create a serious design difficulty.

1.   Analysis

In order to consider the difficulties involved in dissipating heat from an SE climber, we will look at a specific numerical example. For that example, we will look at the design for a construction climber that was spelled out by Brad Edwards [1]. He assumed:

1. 12 m^2 of photovoltaics. (4 m diameter)
2. 100 kW of power produced at 59% conversion efficiency. (169 kW of incident power)
3. 100 kW of motors
4. 900 kg total climber mass (380 kg machine with 520 kg load capacity)

The highest efficiency photovoltaics operate at monochromatic frequencies, as do the lasers that would power them. In order to operate at the highest efficiency, anti-reflective coatings are applied to the surface of the photovoltaic cells to capture the most energy possible. Anti-reflection coatings can be 99.9% efficient at a single wavelength with a fixed angle of incidence. Those anti-reflective coatings would make the surface of the photo cells close to a perfect black body at the desired frequency – absorbing almost all incident laser energy. That means that any laser energy that was not converted into electrical energy would be converted into heat in the photovoltaic cells.

Therefore, the above climber would have to continually dissipate about 69 kW of heat from its photo cells.

Photovoltaics are most efficient near room temperature and decrease in efficiency as the temperature rises. This is because as the temperature increases the band gap of the semiconductor decreases, resulting in a lower junction voltage, and hence a lower output voltage. They can produce 30% less power at 100° C. With broadband solar cells, the lower voltage is partially offset by an increase in current, due to the fact that at higher temperatures, lower frequency photons can start pushing electrons across the smaller gap. On the other hand, with monochromatic photovoltaics, there is no increase in current with higher temperatures, meaning that their loss of efficiency is even worse when they get hotter.

Although the ultimate maximum efficiency of monochromatic photovoltaics is much higher than the 59% assumed by Brad Edwards, we will stick with that number due to the significantly lower efficiency that comes from operating at higher temperatures. We will assume that our climber photovoltaics will need to operate at 100° C. As the rest of this paper will show, such a high temperature is needed to be able to dissipate the heat. However, any higher temperature would probably not be reasonable for the photovoltaics.

At 100° C, the black body radiation of a perfect emitter is 1.1 kW/m², but no body is a perfect emitter. However, there are magnesium oxide paints that have an emissivity ε, of .9 or 90% of the maximum. That paint also has an absorptivity, α, of only .09, meaning that it only absorbs 9% of the incident radiation. That gives an α/ϵ of .1.

With incident solar radiation in space at 1353 W/m², that means that only 122 W/m² would be absorbed onto our surface in direct sunlight. Because the surface would not be in direct sunlight all the time and due to the angles of incidence, we will assume an average of 50 W/m² absorbed over the entire day. Our surface would emit 990 W/m², giving an average net outflow of power of 940 W/m² from a radiator surface at 100° C.

Therefore, to dissipate 69 kW of heat we would need:

$$\frac{69,000 \; W}{940 \; W/m^2} = 73.4 \;\; m^2 \text{ of surface area}$$

If we wanted to operate the photovoltaics at a lower temperature than 100° C, then the required surface area would be much larger.

Initially, one would think that with the photovoltaic array, we would get two surfaces to radiate heat – the top and the bottom of the photo cells. The side away from the laser irradiation could definitely be painted with a coating to give it a very low α/ϵ, making it a good radiator. Unfortunately, the photovoltaic side has been optimized to have a high absorptivity in order to increase its efficiency.

The most efficient monochromatic photovoltaics are made of gallium arsenide, but GaAs has a very low emissivity at infrared frequencies [2] where the heat would be radiated. Therefore with a high α and low ϵ, GaAs photovoltaics would not be able to radiate away any significant net heat from the active side. In fact, that side of the photovoltaics may well be a net heat source from absorbed sunlight, which the other side would have to radiate away.

As a result, we can only count on one side of the photovoltaic array as a radiative surface. That would mean that we would need 73.4 m² of photovoltaic arrays to radiate away the heat, which is over 6 times as much as the original design called for. That would imply a photovoltaic array almost ten meters in diameter, which would be extremely massive and unwieldy.

Instead of increasing the size of the photovoltaic array, we could consider what would be required to add heat sinks to the original 4 m diameter design. However, we would still need 73.4 m² of effective radiative surface.

Here on earth we are accustomed to seeing heat sinks that have a large amount of surface area in a small volume, for example in an automobile radiator. However, almost all the heat that is dissipated from an automobile radiator is through conduction and convection via the airflow. As radiative dissipation is the only means of losing heat in space, high surface area, low volume radiators cannot work there. In space, the radiative surfaces have to be largely isolated from each other so that radiation from one area does not significantly impinge upon another, merely transferring heat from one point to another.

Convoluting the surface of a space-based heat sink would result in a larger surface area, but it would lower the radiative efficiency of that surface. Therefore, an increase in total surface area would be required to dissipate the heat. Obviously, there would be a point at which additional convolutions in the heat sink surface would give no additional benefit, but just add more weight.

A possible heat sink configuration with long fins that could be placed on top of the photovoltaics is shown below in Fig. 1.

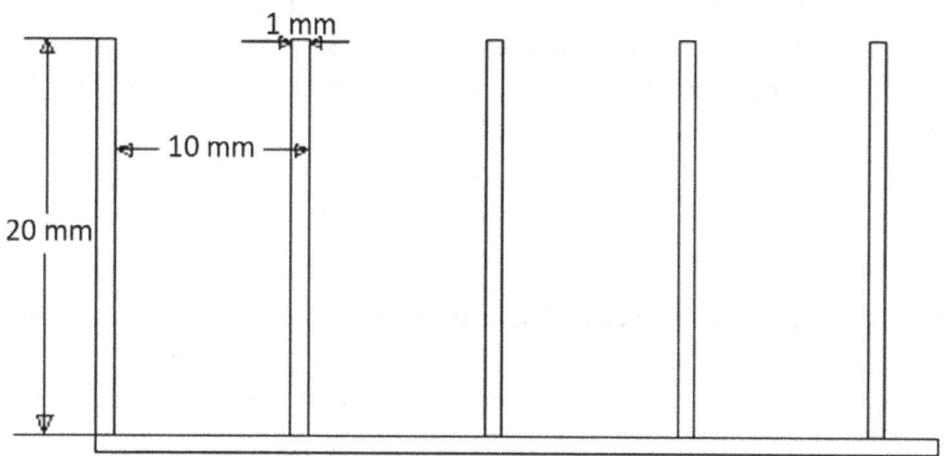

Figure 1. Heat sink dimensions in a cross section

Such a heat sink covering the entire photovoltaic array would increase the surface area by a factor of five. However its effective radiative area would be considerably less than five because photons emitted near the bottom of the fins would typically be reflected multiple times before they could escape, with a chance of absorption each time. It is doubtful that the effective radiative area for this heat sink would be more than 3 or 3.5 times the surface area of the original photovoltaic array. Adding more fins or extending the length of the fins would not increase the effective radiative area. Plus, the heat sink as shown would have a mass of 595 kg, and it still would only dissipate about half the heat.

The bottom line is that there is no possible way that 12 m^2 of photovoltaics could have the equivalent of 73.4 m^2 of effective radiative heat sink attached directly behind it. In order to dissipate the 69 kW of heat, much of that heat would have to be conducted away from the photovoltaic array to a larger radiative array located somewhere else. Ignoring the obvious mass penalties associated with such a plan, let us review what would be involved in conducting heat away to a secondary heat sink.

Let us assume that we could effectively radiate away about half the 69 kW of heat from the heat sinks located behind the photovoltaic array. Then we would only need to conduct about 35 kW of heat away to another location. We will assume that the other location is another heat sink array, located 1 m away.

To conduct heat from one location to another, we have the following relationship:

$$Watts\ conducted = \frac{TkA}{l}$$

$Where\ T\ is\ the\ temperature\ difference\ in\ Kelvins,$
$k\ is\ the\ thermal\ conductivity\ of\ the\ material,$
$A\ is\ the\ cross\ sectional\ area,$
$and\ l\ is\ the\ distance\ the\ heat\ has\ to\ travel.$

Let us analyze the possibility of conducting the heat through a large aluminum plate connecting the two heat sinks. Aluminum typically has a thermal conductivity of about 200 W·K^{-1}·m^{-1}. With a plate 3 meters wide and 5 cm thick, assuming a temperature drop of 30 K from the photovoltaics to the secondary heat sink, and conservatively assuming only 1 m of heat travel, we would have:

$$W = \frac{30(200).15}{1} = 900\ W$$

That is only about 2.5% of the amount of heat that we would need to conduct, yet the plate would have a mass of over 3000 kg in order to collect heat from the array and disperse it to the secondary heat sink. Even if the plate were a pure diamond slab, it would still only conduct 25% of the required heat.

The bottom line is that conducting heat away is not a viable option due to the high power rates, the low temperature gradient, the relatively low cross sections, and the relatively long distances required. A liquid cooling process could conduct heat away to another heat sink much more efficiently than a conductive slab, but it would still add a significant mass in addition to the mass of the secondary heat sink.

With the photovoltaics themselves, heat conduction is not an issue, simply because the heat conduction cross section is so large, and the distance between the sides is so small. As a result, there would only be a negligible temperature drop from one side of a photo cell to the other.

Therefore, the only practical method to dissipate the 69 kW of heat would be to have a 10m diameter photovoltaic array. The other option would be to operate the climber on 1/6th the power level that it was originally designed for, which would increase the construction time of an SE and decrease the throughput by a factor of six. However, the mass of a 10 m diameter photovoltaic array and its required structural supports would vastly decrease the load carrying capacity of the climber also.

In addition to the heat from the photovoltaic array, there is also a considerable amount of heat generated on the climber from the power conversion electronics, the motors, the gearing mechanisms, and the rolling friction and heat of flexure of the wheels. We will estimate that at least 10 kW of heat would be generated in these processes.

Unlike the photovoltaics, this heat is not distributed diffusely over a large area, but instead is generated in relatively compact volumes. Brad Edwards' climber design called for liquid cooled motors, so we will assume that we can also use the same liquid cooling to take heat away from the electronics, gearing, and wheels. Let us review what would be involved in that liquid cooling process.

Liquid cooling is a much more effective process for transferring heat over large distances than is conduction. It is also a very efficient process for extracting heat from concentrated areas. However, the process of dispersing that heat throughout a large radiative array is a little more complicated.

With the photovoltaics, we assumed a maximum operating temperature of 100° C, which gave us a net power radiated of 940 W/m^2 from the radiative surface. With the motors, wheels, gears, and possibly the electronics, we might be able to get a higher operating temperature, and therefore a higher radiative power, even with the inevitable temperature drops associated with the liquid heat exchangers. We will assume that we can radiate a net power of 1000 W/m^2 from the radiator of our liquid cooled system. That means that we would need 10 m^2 of effective surface area.

Because of the laser tracking accuracy we would want any additional heat sink to be in the shadow of the photovoltaic array. Otherwise it would be heated by any laser "spill-over" from the power beam. Also, we would not want it to interfere with the effectiveness of the photovoltaic radiator.

A design that could accomplish those goals would be a 3 m diameter aluminum cylinder, .8 m in height, suspended above the plane of the photovoltaics. It would be 4 mm thick. This design would actually give us more than 10 m^2 of total surface area, counting the inside and outside of the cylinder. However, we would need that extra surface area because of the interference with the photovoltaic heat sink, and the equipment that would be inside the cylinder.

With 10 kW of heat input, the temperature of 1 liter of water only increases 2.4° C per second. With a water flow rate of only 20 L/min, the water temperature would only rise 7.2° C from the heat. Therefore, we could extract heat without a sizable temperature drop from the equipment to the water.

If we wanted a temperature drop of less than 20° C from the liquid to the coolest parts of the radiative heat sink, we could do that with 50 tubes of 1 cm diameter aluminum passing up and down the inside surface of the cylinder at about 20 cm spacing.

That cooling system for the machinery, including the aluminum cylinder and all the necessary tubing, would add a little over 100 kg to the weight of the climber. Those 100 kg would still be a significant load, however, as the original design only had 380 kg of empty weight to begin with.

2.   Conclusion

Although the above study only looked at the design of the smallest construction climber that Brad Edwards envisioned, the same principles would give gloomy results for larger climbers. A 20 ton climber would need 1MW or more of heat dissipation, which would imply more than 1000 m^2 of radiating area. Such a large heat sink would be incredibly unwieldy.

After all the heat loading issues are taken into account, doubt is cast on the practicality of using laser powered climbers on a space elevator.

References

[1] Edwards, B., Westling, E., THE SPACE ELEVATOR A revolutionary Earth-to-space transportation
      system, Houston, Texas, BC Edwards, 2003, p. 54

[2] Aleksandrov, S.E., et al, "RADIATION THERMOMETER CONFIGURED FOR GaAs MOLECULAR BEAM EPITAXY", Proc. Of SPIE, Vol. 5066, p. 165, 2003

ENHANCING SPACE ELEVATOR SAFETY BY ACTIVE DEBRIS REMOVAL

Jerome Pearson[1], Eugene Levin[1], and Joseph Carroll[2]

[1]STAR, Inc., 3213 Carmel Bay Dr., Ste 200, Mt. Pleasant, SC 29466 USA, Email: jp@star-tech-inc.com
[2]Tether Applications, Inc., 1813 Gotham Street, Chula Vista, CA 91913, USA, Email: tether@cox.net

ABSTRACT: *We propose a low-cost method for improving space elevator safety by removing space debris. The ElectroDynamic Debris Eliminator (EDDE) can affordably remove nearly all the 2500 objects of more than 2 kg that are now in low Earth orbit. They have more than 99% of the total mass, collision area, and debris-generation potential in LEO. These are the debris objects that most threaten the safety of a space elevator. EDDE is a propellantless vehicle that reacts against the Earth's magnetic field. EDDE can climb about 200 km/day, decay at 1200 km/day, and change orbit plane at 1.5°/day. No other electric vehicle can match these rates, much less sustain them for years. Capture uses lightweight expendable nets and real-time man-in-the-loop control. After capture, EDDE can drag its payload down to release it and the net into a short-lived orbit safely below ISS, sling the debris object into controlled reentry, or can even capture upper stages and re-cycle them for aluminum space construction.*

A dozen compact, 100-kg EDDEs could remove nearly all 2200 tons of large LEO orbital debris in 7 years, clearing the way for space elevator construction. Smaller objects could be removed later. Coupled with a Hall thruster, EDDE could also remove dangerous space debris in medium Earth orbits, eliminating the need for the space elevator to actively maneuver to avoid debris. The estimated overall program cost is on the order of $100 per kilogram of debris removed. A dynamic simulation shows the removal operation.

1. Introduction

Space debris from discarded upper stages, dead satellites, and assorted pieces from staging and tank explosions has been growing since the beginning of the space age. This has increased the risk to active satellites, and the need for avoidance maneuvering. These thousands of pieces of space junk in Earth orbit pose risks to our space assets such as communication and navigation satellites, environmental monitoring satellites, the Hubble Space Telescope and the International Space Station (ISS). More importantly, they pose a risk to the astronauts who work outside the space station or who repair satellites, as the space shuttle Atlantis astronauts did for Hubble last year. In addition to the Hubble's bad camera and failing gyros, its solar array had a hole in it the size of a .22-caliber bullet. Figure 1 is a depiction of the tracked objects over 2 kg intersecting the space elevator orbital plane in low Earth orbit (LEO). The space elevator ribbon is expected to be struck about every 4 months somewhere in LEO [1].

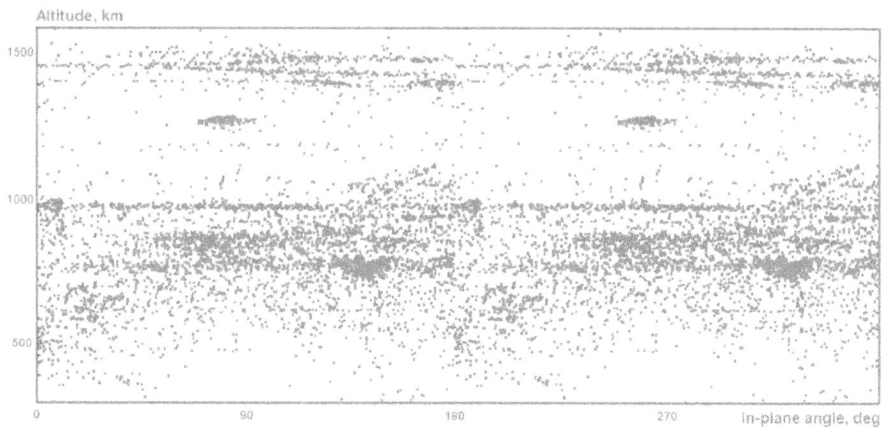

Figure 1. Tracked Objects in LEO Crossing the Orbital Plane of the Space Elevator

The U.S. Strategic Command tracks about 20,000 pieces of debris larger than 10 cm, and there are an estimated 300,000 pieces larger than 1 cm (about the size of a .38-caliber bullet) [2]. At typical speeds of 5-10 km/sec, a piece of debris will go right through a satellite, leaving a trail of destruction. The ISS and the space shuttles occasionally have to adjust their orbits to avoid known pieces of space debris, and the inevitable future collisions of large debris objects will create thousands of new "bullets" that will increase the danger exponentially in time.

There have already been four recorded collisions with space debris. In December 1991, the Russian satellite Cosmos 1934 collided with debris from Cosmos 926. In July 1996, the French satellite CERISE was hit by a debris object. The next identified event was the January 2005 collision between THOR BURNER and debris from a Chinese Long March rocket. Most recently, on February 10, 2009, the Iridium 33 satellite was destroyed in a collision with Cosmos 1421.

There have also been deliberate acts that caused space debris, most notably the Chinese anti-satellite weapon test of January 11, 2007 [3], which destroyed an aging Chinese weather satellite, Fenyung 1-C, at 865 km altitude. The collision caused a debris cloud of more than 2300 tracked objects and perhaps 35,000 bits larger than 1 cm. The debris objects ranged from 200 km to 3850 km in altitude, and endangered all satellites in LEO. The U.S. also tested anti-satellite weapons in 1985 and in early 2008, the latter to destroy a malfunctioning spy satellite, but these were at lower altitudes and the debris was shorter-lived, especially the 2008 test, which occurred at 247 km altitude. Because of the international outcry over the Chinese test, no other tests like these are expected.

2.   Keeping Track of Debris

The U. S. Strategic Command keeps track of the 20,000 catalog debris objects and the 800 active satellites, calculates potential collisions, and issues warnings to satellite operators. Each day they produce 800 "conjunction analyses," about one for every active satellite. Many satellites can maneuver out of the way of debris when a near approach is predicted. However, STRATCOM does not have the resources to predict every potential conjunction, and there was no warning issued on the Iridium satellite in 2009.

The NASA Orbital Debris Program Office at the Johnson Space Center in Houston studies space debris and formulates rules to limit debris creation. These rules include eliminating throwaway bolts and latches when spacecraft are placed in orbit, venting fluid tanks to prevent explosions, and requiring that satellites re-enter the atmosphere within 25 years after their missions are completed. But the office director, Nicholas Johnson, says that unless we begin removing existing debris from orbit, the inevitable collisions involving objects like 8-ton rocket bodies and 5-ton dead satellites will create tens of thousands of new pieces of debris, resulting in a "debris runaway" that would make LEO unusable for hundreds of years. This so-called "Kessler Syndrome," forecast by Don Kessler in 1978 [4], would make it impossible to build or operate the space elevator.

Up until 2009, the dangers of space debris were generally ignored under the "big sky" view that space is very empty. But the loss of the operating Iridium 33 satellite changed that. Since then, there have been Congressional hearings and international conferences discussing the problems of space debris, how to reduce the risks, and whether we can afford it. This year the proposed new NASA Office of the Chief Technologist is soliciting solutions to the debris problem.

A December 2009 conference sponsored by NASA and DARPA (the Defense Advanced Research Projects Agency), featured many proposed solutions, including large orbiting shields to catch small debris, ground-based lasers to ablate the front side of debris to increase drag, and active spacecraft to capture the larger debris items and drag them down to atmospheric entry [5].

3.   The EDDE Solution

The most near-term and technically advanced method presented was a roving space vehicle that can capture LEO debris objects in nets and drag them down safely out of the space lanes. EDDE, the ElectroDynamic Debris Eliminator, is the first space vehicle that can remove all the large debris from LEO at reasonable cost [6]. A schematic of EDDE is shown in Figure 2. It consists of a long conductor, solar arrays, electron collectors and emitters, and net managers at each end to deploy large, lightweight nets to snare debris objects.

EDDE is a new kind of space vehicle [7]. It is not a conventional rocket that accelerates a payload by throwing propellant mass in the opposite direction. EDDE is an electric motor/generator in space that accelerates by reacting against the Earth's magnetic field, and is therefore propellantless. This means that it is not limited by the Tsiolkovsky rocket equation, and can produce enormous delta-Vs of hundreds of km/sec over its operational lifetime. An EDDE vehicle equipped with solar panels for power and expendable capture nets could safely remove from orbit its own mass in debris each day on average.

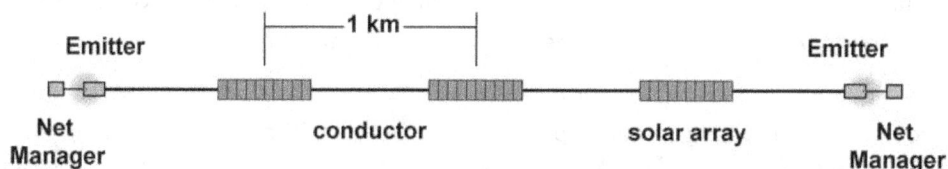

Figure 2. Representative Sections of the 10-km-Long EDDE Vehicle

The principle of operation of an EDDE vehicle is shown in Figure 3. The vehicle is in low Earth orbit, moving in the Earth's dipole magnetic field and surrounded by the ionized plasma from the solar wind that is trapped in the ionosphere. Solar arrays generate an electric current that is driven through the long conductor; the magnetic field induces a Lorentz force on the conductor that is proportional to its length, the current, and the local strength and direction of the magnetic field. Electrons are collected from the plasma near one end of the bare conductor, and are ejected by an electron emitter at the other end. The current loop is completed through the plasma.

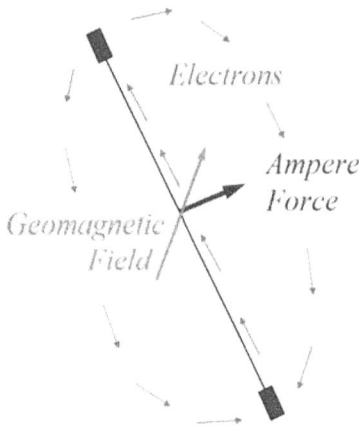

Figure 3. Propellantless Propulsion

This propellantless propulsion was demonstrated in orbit by NASA Johnson on their Plasma Motor Generator experiment. The average thrust going down can be considerably higher than that going up, because energy is being extracted from the orbital motion.

The EDDE vehicle is a very unusual spacecraft; it is two micro-satellite end bodies connected by multiple 1-km-long segments of reinforced aluminum ribbon conductor just 30 mm wide and 38 microns thick, as shown in Figure 4. The bare aluminum conductor is an electron collector, and each end body contains an electron emitter. Solar arrays are distributed along the length, and the entire structure rotates slowly end over end to maintain tension and stability, a key patented advance in making its high performance possible [8] [9] [10]. The rotation rate is typically a few revs per orbit, and can be controlled by reversing the currents in different sections of the conductor. The rotation plane is also controlled.

Because there are many units of each element in the electrical circuit, even if EDDE were cut in two by a meteoroid, each end could still function as an independent satellite and can safely deorbit itself. For debris removal, each end body is equipped with a net manager that carries about 100 Kevlar nets of 50 g each. To catch a debris object, a net is extended by the rotational force as the EDDE end approaches the target at a few meters per second. The net snares the target, and EDDE actively damps out the dynamics, even if the object is spinning or tumbling up to about 1 rpm. Most debris objects are rotating much slower than this because of the eddy-current damping of their aluminum structure.

A further advantage of the EDDE propellantless spacecraft is that it folds up very compactly. Despite deploying to 10 km long, it folds up into a compact box 60 cm square and 30 cm deep. This allows two EDDEs to fit into one of the secondary payload slots of the Boeing Delta 4 or Lockheed Atlas 5 ESPA ring. It can also be launched as a secondary payload on the Orbital Sciences Pegasus air-launched vehicle, and the new SpaceX Falcon 1 and Falcon 9. If there is some payload margin for the launch vehicle, then there is little additional cost to launch EDDE vehicles piggyback. One EDDE vehicle can fit into each secondary payload slot, leaving room for several nanosatellites that EDDE can carry to custom orbits after the primary payload is released, as shown in Figure 4.

For typical values of a few kilowatts of electric power from large, thin-film solar cells, a few amps of current and a 10-kilometer-long conductor, the force is typically half a newton. This is a very small force compared with typical rockets, but its advantage is that it operates continually, orbit after orbit, gradually changing the orbital elements. The result is a low-thrust system that changes orbits slowly, but has very high capability for very large orbit changes. The thrust is several times that of the ion rocket that drove the NASA Deep Space 1 probe to Comet Borrelly in 2001. Reversing the current at the right times around each spin and each orbit allows any desired combination of coordinated changes in all 6 orbit elements. The force can also be used to change the orientation of the conductor by changing the EDDE rotation rate and plane.

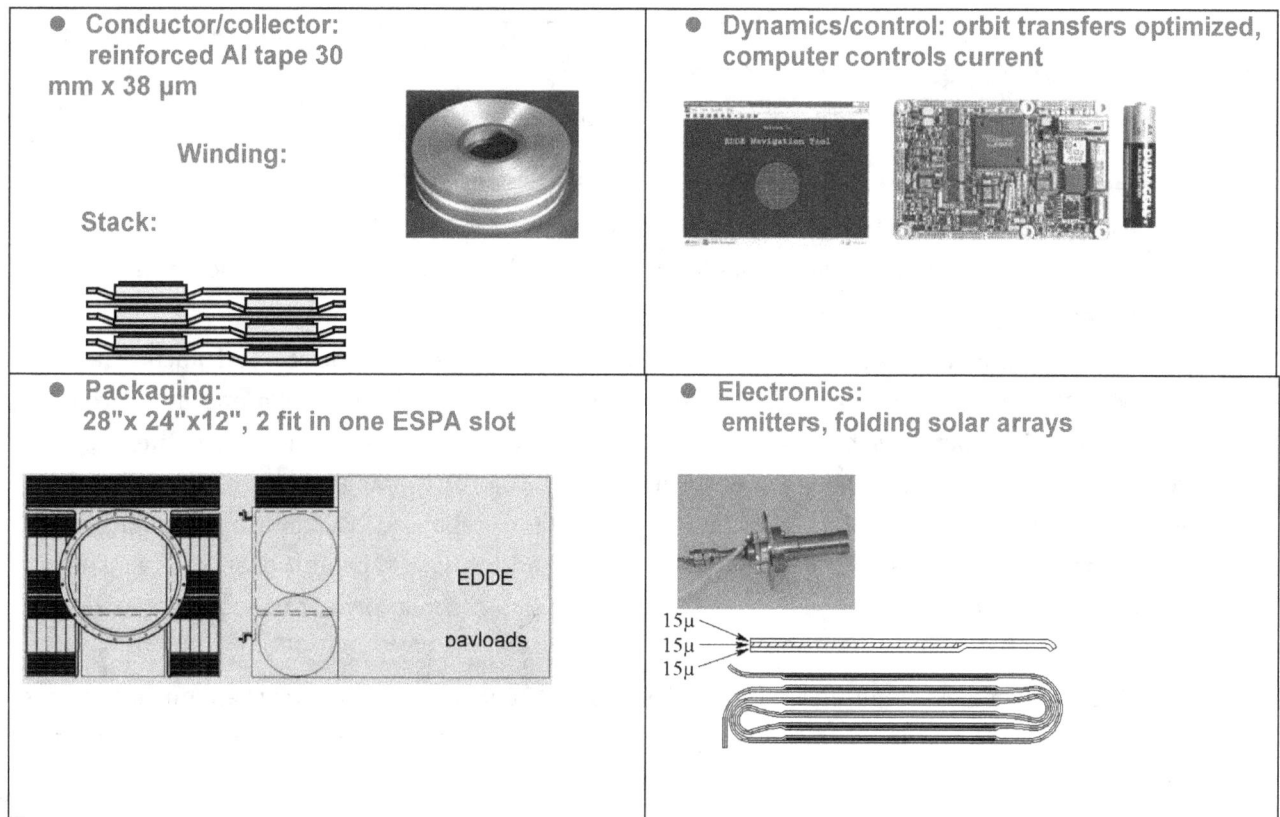

Figure 4. The EDDE Conductor, Dynamics and Control, Packaging, and Electronics

By using lightweight solar arrays, a reinforced aluminum ribbon conductor, and hollow cathodes at each end to run reversible currents, a typical EDDE spacecraft produces about 7 kW of power and weighs 100 kg, and can make large changes in its orbit in a fairly short time. Figure 5 shows typical rates of change of inclination, node, and altitude of the EDDE orbit [11]. The deboost rate can be much higher than shown, because additional energy is extracted from the orbital motion through emf.

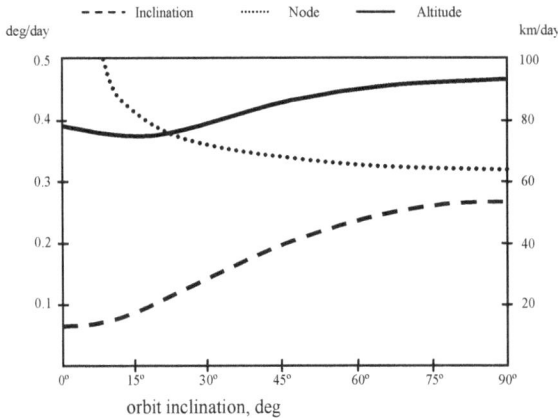

Figure 5. EDDE Orbit Transfer Performance

These rates are possible over altitudes of about 300 km to 1000 km, and are reduced at higher altitudes by lower magnetic field strength and plasma density. A bare EDDE vehicle without a payload could go from the International Space Station 51.6° inclination orbit to 90°-inclination polar orbit in about 4 weeks, a delta-V of more than 6 km/sec.

Using conventional rockets for space debris removal is extremely difficult. To launch a satellite into low Earth orbit, it must be given a velocity of 7 or 8 km/sec. With chemical propellants, even our best launch vehicles put only about 4% of the total launch mass into orbit. But to change the orbit of a satellite already in orbit can require even higher velocities. For example, to move a satellite from equatorial to polar orbit takes 1.4 times the orbital velocity, or about 10-11 km/sec. It would actually be easier to launch another satellite from the ground than to make this orbit change! Launching a chemical rocket from the ground to remove the debris, each piece in its own orbit, would be extremely expensive.

The enormous advantage that the propellantless EDDE vehicle has over conventional rockets is shown in Table 1, which compares different propulsion systems in performing the task of removing the 2465 objects in LEO weighing over 2 kg.

Table 1. Propulsion System Requirements for Debris Removal

Propulsion System	Isp, sec	Number of Vehicles	Total Mass in Orbit
Bipropellant	300	900	800 tons
NH_3 Arcjet	800	300	250 tons
Ion Rocket	3,000	120	65 tons
VASIMR	10,000	30	25 tons
EDDE	----	**12**	**1 ton**

A typical bipropellant chemical rocket might have specific impulse of 300 seconds, and the table shows that this task would require 900 vehicles weighing 800 tons. Higher-Isp systems include arc jets, ion rockets, and the recently-tested Variable Specific Impulse Magneto-plasma Rocket (VASIMR) championed by former NASA astronaut Franklin Chang-Diaz of Ad Astra [12]. These systems also require higher power.

But even VASIMR would require 25 tons in orbit to remove all the debris, more than 20 times the mass of 12 EDDEs, a little over 1 ton. Twelve EDDEs could remove all 2465 objects, a total of 2166 tons, in less than 7 years.

4. Removing LEO Debris

The total LEO debris mass of the 2465 largest objects is 2166 tons. Many are in orbits of 81°-83° or 70°-74° inclination. There are also many old satellites in "sun-synchronous" orbits that allow them to pass over the same spot on the Earth at the same time each day; these are typically 400-800 km in altitude and 97°-99° in inclination, slightly retrograde.

The EDDE performance in doing this task of removing all the large debris from LEO has been analyzed in detail by Star Technology and Research, and Tether Applications, the two small companies who developed the EDDE concept and vehicle. Table 2 shows the typical removal operations that the EDDE vehicles would perform in dragging each LEO debris object down to an altitude 20 km below ISS, reducing its orbit life to a few months. It takes 10 days to remove an average debris object of 1000 kg from an average altitude of 800 km, which means that on average each 100-kg EDDE vehicle can remove its own mass in debris every day. EDDE could also provide targeted re-entry by using a ballast mass of another piece of debris to throw the debris into the southern Pacific Ocean [13].

A simulation of the debris removal using 12 EDDE vehicles in 6.7 years is available at: http://www.star-tech-inc.com/id121.html. This simulation shows the history of the space debris buildup from 1959 to 2010, and snapshots of its removal by EDDE. The full simulation is shown in a link on the same web page, which also has links to presentations given at the DARPA/NASA International debris removal conference in December 2009.

Table 2. Typical EDDE Timelines with 1-Ton Objects

Operation	Days	Typical Parameters
Phase to next target	0.4	400 km Δaltitude, ½ orbit average
Climb and tune orbit	2.9	200 km/day, +20% for plane change
Approach and capture	0.5	6-8 orbits at 800 km average altitude
Deboost and release	6.2	90 km/day, +20% margin
Total per target	**10**	**1000 kg object, 800 km To 330 km altitude**

5. Debris Removal Options

There are several options in disposing of debris objects. The objects can be lowered to 330-km orbit for rapid natural decay or released at 330-km with a smart drag device for controlled reentry. They can be released to targeted reentry using momentum exchange between EDDE and debris object. This kind of maneuver has been demonstrated in orbit by SEDS-1 (US) and YES-2 (ESA/Russia) tether experiments.

Another option that is gaining attention is recycling. There is enough aluminum in old upper stages at high inclinations to build very large new structures in orbit that can never be launched from Earth. This way, not only will the debris be removed, but valuable orbital resources will be utilized for future projects. We could use space construction from aluminum upper stages to create the large complex at the geostationary hub of the space elevator, far larger than could be carried up the ribbon by conventional climbers under 1 g.

6. Other Removal Methods

There are other methods for debris removal using electrodynamic tethers, but they are far less effective and far more risky than EDDE. It has been suggested that rockets could be used in a single orbit inclination to attach drag devices such as balloons or passive electrodynamic tethers to drag the debris down.

First of all, any debris removal scheme that involves rockets with chemical propulsion will be much more expensive by itself, but there is also another problem. These devices do not actively control the debris deorbit process for collision avoidance, but they have much larger collision cross-sections compared to the debris itself, and they will add to the collision risk during the long time they take to de-orbit. Using passive electrodynamic tethers, for example, would require having multi-kilometer tethers on hundreds of objects over years as they slowly spiral down to re-entry, which would result in a huge additional collision risk, especially to the ISS.

There are proposals for de-orbit modules that would be launched with new satellites [14], but these modules would not address the existing debris that must be removed. Also, they are passive and are at risk of collision during the de-orbit process. The advantages of EDDE are that it handles existing debris, removes debris objects quickly, each object within days, and actively avoids all tracked objects while dragging debris to disposal.

7. Tether Technology Maturation

Several successful spaceflights have demonstrated tether deployment and operation in orbit. The SEDS-1 and SEDS-2 flights by NASA Marshall deployed 20-km-long braided Spectra tethers, and SEDS-1 sent a 26-kg end-mass into a controlled entry. The SEDS hardware mounted on the Delta II rocket is shown in Figure 6.

Figure 6. SEDS Hardware on Delta II

The PMG, (Plasma Motor Generator) flight by NASA Johnson demonstrated motor/generator operation with a 500-m copper wire and a hollow cathode, which is enabling for the EDDE vehicle. A view of the flight winding for the PMG conductor is shown in Figure 7.

The TiPS (Tether Physics and Survivability Experiment) by the Naval Research Laboratory demonstrated a long lifetime for a 2-mm by 4-km tether connecting two end masses.

The tethers and deployers for SEDS-1, SEDS-2, PMG, and TiPS were designed and fabricated by Joe Carroll of Tether Applications (www.tetherapplications.com).

Figure 7. PMG Conductor Winding

Joe Carroll also designed and fabricated electrodynamic tethers and deployers for ProSEDS (Propulsive Small Expendable Deployer System) and METS (Mir Electrodynamic Tether System). ProSEDS was built by NASA Marshall to demonstrate de-orbiting of Delta II upper stage using a 5 km electrodynamic tether. It was scheduled for launch in August 2000, but canceled after the Columbia accident. METS was built to keep Mir in orbit without fuel re-supply using a 7.5 km electrodynamic tether. It was scheduled for launch in early 2001, but canceled due to the decision to de-orbit Mir. The work on both projects has greatly advanced technologies crucial for electrodynamic tethers.

Figure 8. TiPS End Masses and Tether Winding

A new electrodynamic tether system, TEPCE (Tether Electrodynamics Propulsion CubeSat Experiment), is being developed by NRL. It will be a 3-unit CubeSat demonstration of emission, collection, and electrodynamic propulsion [15] planned for 2011. The two end pieces will be connected by a 1-km conducting tether stowed in the center cube. The conductive tether design and stacer-driven deployment technique were proposed by Joe Carroll. Figure 9 shows a TEPCE prototype before a recent deployment test.

Figure 9. TEPCE Hardware for Deployment Test

Following the TEPCE flight and a possible follow-on, the EDDE program is aimed at a Mini-EDDE spaceflight of a scaled-down 50-kg vehicle 2-3 km long that will demonstrate large orbit changes and rendezvous. This will be followed by a mission-capable EDDE that can fly piggyback on any flight with a 100-kg payload margin. It will demonstrate the capture and de-orbiting of an inactive US object using a deployable net.

## 8.	Space Traffic Management

EDDE is the first example of a new class of space vehicles that can move all over LEO performing a variety of jobs such as removing orbital debris and delivering payloads. But in order to perform its functions and move freely among other objects in LEO, constantly changing its orbital elements, EDDE needs a way to coordinate its flight plans with the plans of other spacecraft operators. A centralized flight coordination service such as offered by Center for Space Standards and Innovation (CSSI) can support EDDE operations in the future.

Additional work would needs to be done on the tracking side. EDDE has an unusual radar signature, and consists of several separate objects moving in non-Keplerian orbits. New tracking algorithms will need to be developed to calculate the orbit of the center of mass of the EDDE vehicle, and to predict future positions by keeping track of the changing orbit from the continuous electrodynamic thrusting.

## 9.	Operational Questions

A number of operational issues will have to be resolved for EDDE to be able to perform its debris collection function, including agreements to capture debris objects, space object registration transfer or some other arrangement before handling any foreign-owned objects, if required, insurance for EDDE operator and debris owner, agreement on disposal or recycling methods, safety requirements on debris capture and removal.

## 10.	EDDE Applications

EDDE can be used for a variety of useful purposes other than debris removal. EDDE can deliver payloads to custom orbits, deliver fuel to operational satellites, deliver service modules to satellites, deliver satellites to ISS for service, move satellites to new orbits, inspect failed satellites, and monitor space weather all over LEO. Multiple EDDE vehicles in different orbits could provide real-time maps of the ionosphere, keeping track of "space weather," which affects satellite communication, and could also record the effects of solar flares and proton events on the Sun, which are dangerous to satellites and crew.

Perhaps more importantly, EDDE could be used in conjunction with astronauts on the International Space Station to repair and refurbish aged or failed satellites. EDDE could capture them with a gentle latching mechanism, avoiding damage to their solar arrays, and bring them to the ISS for test and evaluation. After they are fixed, perhaps with new components brought to the ISS from Earth, EDDE can take them back to their original (or new) orbits for continued operation. There have been billion-dollar satellites that failed soon after launch, and such on-orbit repair operations could be a very valuable part of full-scale ISS operations.

Finally, EDDE propulsion systems could be used to maintain the ISS in orbit without expending rocket fuel, and even control it to avoid collisions with a space elevator under construction or in operation. Similarly, EDDE propulsion could be used to move the Hubble Space Telescope into a different orbit, and prevent it from colliding with the space elevator.

11.   Conclusions and Recommendations

The EDDE vehicle is based largely on concepts already proven in flight, mostly on projects in which EDDE team members played key roles. Some of EDDE's novel aspects are planned for test as part of NRL's TEPCE experiment next year, and others are being considered for a potential TEPCE-II test. We plan to mature all other novel aspects of EDDE under current SBIR and follow-on funding. We hope to be ready for an integrated ~3 km "Mini-EDDE" flight test within 4 years. This test would use full-scale EDDE components, but fewer of them than in a full 10 km EDDE.

Starting with next year's TEPCE test, this sequence of flight tests will validate EDDE's persistent maneuvering capability and allow extensive testing and refinement of EDDE components and software. Iterative refinement of software for control, rendezvous, and active avoidance of other tracked objects will also allow TEPCE and EDDE to assist the testing of upgraded space tracking and traffic management capabilities.

After these test missions, EDDE should be ready to begin routine commercial operation, delivering and recovering satellites and removing debris from LEO, for both U.S. and foreign customers, government and commercial. It seems unlikely that plausible improvements in alternative concepts can make any of them competitive with EDDE, because the same wholesale debris removal that requires about 1 ton of EDDE vehicles would require 25 to 800 tons of vehicles using other propulsion methods.

The EDDE system could be an important part of the plan to make the space elevator possible, by reducing the risk of catastrophic failure, and by reducing the number and severity of the inevitable small punctures from trackable debris. That tracking limit is now about 10 cm in dimension, but in the future it will be extended to the detection and tracking even smaller objects, enabling EDDE vehicles to improve space elevator safety even further.

References

[3] Swan, P., Penny, R., and Swan, C., "Space Elevator Survivability—Space Debris Mitigation," Draft Study for 2010 Space Elevator Conference, Redmond, WA, 13-15 August, 2010.

[4] Liou, J.-C., and Johnson, N. L., "Risks in Space from Orbiting Debris," Science Vol.311, pp. 340-341, 20 January 2006.

[5] David, Leonard, "China's Anti-Satellite Test: Worrisome Debris Cloud Circles Earth," Space.com, 2 February 2007. http://www.space.com/news/070202_china_spacedebris.html

[6] Kessler, D. J., and Cour-Palais, B. G., "Kessler Syndrome Collision Frequency of Artificial Satellites: The Creation of a Debris Belt," Journal of Geophysical Research, Vol. 83, No. A6, 1978, pp. 2637-2646.

[7] Klinkrad, Heiner, and Johnson, Nicholas, "Space Debris Environment Remediation Concepts," NASA-DARPA International Conference on Orbital Debris Removal, Chantilly, VA, 8-10 December 2009

[8] Pearson, J., J. Carroll, E. Levin and J. Oldson, "EDDE: ElectroDynamic Debris Eliminator for Active Debris Removal," NASA-DARPA International Conference on Orbital Debris Removal, Chantilly, VA, 8-10 December 2009. http://www.star-tech-inc.com/papers/EDDE_for_Debris_Conference.pdf

[9] Pearson, J, J. Carroll, E. Levin, J. Oldson, & P. Hausgen, "Overview of the ElectroDynamic Delivery Express (EDDE)," paper AIAA-2003-4790, 39th Joint Propulsion Conference, Huntsville, AL, July, 2003. http://www.star-tech-inc.com/papers/edde/edde_2003.pdf

[10] Levin, E. and J. Carroll, "Method and Apparatus for Propulsion and Power Generation Using Spinning Electrodynamic Tethers," US patent 6,942,186, Sept 2005.

[11] Levin, E. and J. Carroll, "Apparatus for Observing and Stabilizing Electrodynamic Tethers," US Patent 6,755,377, June 2004.

[12] Levin, E. and J. Carroll, "Method for Observing and Stabilizing Electrodynamic Tethers," US Patent 6,758,433, July 2004.

[13] Levin, E., Dynamic Analysis of Space Tether Missions, Vol. 126 in Advances in the Astronautical Sciences, American Astronautical Society, San Diego, CA, 2007.

[14] Cassady, Leonard, et al., "VASIMR Technological Advances and First Stage Performance Results," AIAA Paper 2009-5362, 45th Joint Propulsion Conference, Denver, CO, 2-5 August 2009. http://www.adastrarocket.com/aarc/Publications

[15] Carroll, J.A., "Space Transport Development Using Orbital Debris," NIAC Phase I presentation and final report, Oct/Nov 2002.

[16] Voronka, N., et al., Technology Demonstrator of a Standardized Deorbit Module Designed for CubeSat and RocketPod Applications, 19th AIAA/USU Conference on Small Satellites, Paper SSC05-XI-4, Logan, UT, 8-11 August 2005.

[17] Coffey, S., Kelm, B., Hoskins, A., Carroll, J. A., and Levin, E. M., "Tethered Electrodynamic Propulsion CubeSat Experiment (TEPCE)," Air Force Orbital Resources Ionosphere Conference, Dayton, Ohio, 12-14 January 2010.

QUESTIONING THE SPACE ELEVATOR LEGAL RISK MANAGEMENT REGIME

By Romain Loubeyre [1],
romain.loubeyre@u-psud.fr
Student in Space Law, Master Degree,
University Paris-Sud 11, Faculty Jean Monnet,
Paris, France

ABSTRACT: *As an innovative way to access outer space, the Space Elevator system could not be commercially operated without an adapted and efficient risk management regime. Indeed, being the safest and most economical means of transportation that could be used to place satellites into orbital positions or to develop an accessible and cost effective space tourism activity, the Space Elevator project could represent the future of all current space launch activities. However, the transport of diverse kinds of payloads and passengers by this radical new mode of transportation is not addressed by the risk management regime currently applied to typical space launch activities, considering the differences in technologies involved. Moreover, the development or adaptation of a legal regime that deals with the Space Elevator system should be considered a priority, considering the demands and guarantees required by capital investors who will fund the project. Classically, a risk management regime is based on three main steps that should be applied to the Space Elevator project: an identification of the multiple risks threatening the project and their sources, an assessment of the potential severity of loss arising from each risk identified and their probability of occurrence, and finally the implementation of risk mitigation strategies. A basic examination of the first two steps demonstrates that the Space Elevator induces a major decrease of certain risks due to the replacement of rocket propulsion; whereas, the deployment of a permanent structure in outer space brings new risks that will have to be addressed, such as space debris mitigation. By comparing these new risks to "traditional" space launch operations, this study will assess the modifications that should be implemented in order to adapt a risk management regime to the Space Elevator. A successful and efficient adaptation shall utilize three main mechanisms: avoiding risks by respecting adapted and strict norms of security; reducing and sharing the risks by designing an original scheme of liability; and considering the unique risk profile of the Space Elevator system, financing the risk by an insurance system that offers a full and specific coverage to each participant involved in the operation of the Space Elevator.*

1. Introduction

Realizing the Space Elevator project would not only mean a change in the way we access to outer space, but also a revolution of the entire space industry, and by extension to humankind which has become highly dependent on space applications such as satellite telecommunications, positioning and remote sensing. A functional Space Elevator makes a wide array of activities possible, from the shipment of payloads to a highly developed and accessible space tourism industry. Someday our children may be able to spend their holidays on the moon or in space, realizing the dreams of futurists and science fiction authors alike. However, before these dreams can be realized, certain realities must be addressed. Before development or construction can begin, the Space Elevator concept must be transformed into a realistic and economically viable project in order to attract the necessary capital investment. This goal will be achieved in two ways: first, an effective technical development guaranteeing the physical safety of the Space Elevator operation, and second, a complete legal risk management regime, which is the topic of this study.

Classically, legal risk management regimes [2] are designed to protect, legally and economically, each party involved, including third parties, in a contract or transaction from damages that could arise from an activity involving a high level of risk. This protection is offered by predetermining the liabilities of the parties involved and by giving the parties a better understanding of the financial risks arising from any potential incident, which is important in commercial activities involving a high level of risk such as the space launch industry. The legal risk management regime that exists for the normal commercial space launch operations has proven to be effective, preserving the industry even in the face of multiple failures and protecting the parties involved, from the largest launch operators to the smallest subcontractor.

The first question is whether the existing risk management system could be applied to the operation of the Space Elevator- the short answer is no. First, the system and business model is completely different, which affects changes to the risk profile as well. Second, to be as effective as possible, a legal risk management regime should be individually tailored to a specific project in order to be entirely adapted to its specificities, and to best protect the parties with a financial or other interest in the project. However, this does not mean that the entire regime must be revised. Some of the techniques applied to the traditional space launch industry could be applied, perhaps with some modification, to the Space Elevator.

Another aspect of the Space Elevator that should not be forgotten while drafting a legal risk management regime is that in addition to payloads and astronauts, the system might also be used to ship public passengers, creating an accessible and highly developed space tourism industry. In terms of legal risks, Space Elevator tourism requires a risk management regime capable of handling a catastrophic worst scenario, including the loss of tourists' lives and the liabilities that can emerge from such an event. To preserve the economic viability of the Space Elevator industry as well as maintaining a stable level of activity even after such a failure, and adequate risk management regime must be employed. Considering foregoing, the space tourism industry will not be the subject to the following developments.

Because the Space Elevator is at an early stage of technical development, drafting a comprehensive and complete legal risk management regime is not possible, considering the level of detail needed to address all the potential risks the project can encounter. However, analysis of whether or not the techniques developed for the existing space launch industry could be used for the Space Elevator system, even at this stage, will clarify some essential elements necessary to design the final risk management regime. After a brief but essential analysis of the risk profile of the Space Elevator, this study will concentrate on specific issues, especially contractual practices, legal framework on space operations and insurance coverage, discussing for each of these topics the suitability of the current regime for the Space Elevator system and the necessary adaptations.

2. Risk Profile of the Space Elevator System

As the project stands today, the operation of the Space Elevator system instead of rocket-based space launch systems would require multiple changes in the risk profile of the shipment of payloads to outer space.

2.1.   Risks reduced by the Space Elevator system

The combination of sunlight and laser light projected from the ground used as the energy source of the Space Elevator greatly reduces risks of explosion during the travel to outer space, compared with rocket propulsion which uses a tremendous amount of propellant that can sometimes be unstable.

The use of a permanent structure is also more reliable since the production of a launch vehicle for each travel multiplies the risks of a malfunction due to an error in the manufacturing process, which has to be repeated for each launch. Because the production of single-use launchers multiplies the risk of a manufacturing error, a Space Elevator also greatly reduces the product liability risks for subcontractors involved in the production of the system. A permanent and reliable structure also means a decrease in risks of damages to people and property on earth and thus a decrease in the probable maximum loss in third party liability[3], which would limit the cost of third party liability insurance and constitute an incentive for insurers to commit themselves to a new space insurance market (that would have to emerge if the Space Elevator is erected).

While a rocket-based system begins an irreversible process as soon as it has launched, the Space Elevator technology would enable the operator to keep a permanent control over the entire "launch" process, meaning that if a problem is detected while the payload is being shipped, the operator may be able to reverse the process and correct it. In the case of an error during a rocket-based launch, the operator is powerless to do anything. In the best-case scenario, the payload might suffer partial damages and in-orbit insurance will compensate the satellite owner for the damage, while in the worst-case, the launcher might be destroyed and the satellite owner will suffer a total loss of its property and significant consequential damages. This possibility would be diminished or avoided by using the Space Elevator system, which would enable the operator to entirely control the process. Finally, depending of the technical characteristics of the project, the risks of damaging the payload during the process would be greatly diminished by a reduction of the vibration of the structure compared to a rocket-based launcher [4].

2.2.   Risks generated by the Space Elevator System

The main source of risk generated by the Space Elevator system is due to the presence of a permanent structure in outer space, although depending on the final characteristics, the project may create some additional risks. Assuming that the structure will be capable of moving to avoid the trajectory of in-orbit satellites and the largest space debris, the growing number of small debris in circumterrestrial space makes it unlikely that the system will be able to circumvent every piece of space debris that could cause damage to the structure.

Damages to the structure is not itself a source of legal risk, but the cost of the repairs to such a complex structure and the resulting delays in the transit schedule could jeopardize the economic viability of the project. This is a serious concern since the timeframe for repair is an unknown. This risk might be managed by a realistic and flexible schedule, allowing modifications by the operator.

If a payload is damaged during the process of in-orbit placement, this could be a source of contractual liability depending of the content of the contract itself. The operator could be held responsible for damages caused to the payloads during the process and could be obliged to indemnify its clients in such an event. The probability that such an event could occur might necessitate a cross-waiver of liability modeled on existing space-launch agreements.

Because of the capital outlay required to build a Space Elevator, the risk that the structure could be totally destroyed also must be seriously considered. [5] The overall conclusion that can be drawn from the information at our disposal is that using the Space Elevator instead of rocket-based launch systems should substantively reduce the risk in an activity that is as highly hazardous and inherently dangerous as space-launch. This risk reduction could potentially transform the entire space industry by making it more accessible, safe, economically profitable and socially beneficial.

First however, a complete, adapted, and reactive legal risk management regime must be designed to ensure the viability and profitability of this new space industry. As previously noted, such analysis cannot be performed until the complete details of the project are known, which is why this article will only study which of the current techniques used for rocket-based launches could be implemented in the Space Elevator legal risk management regime.

3.   Risk Management by Contractual Practices

Managing legal risks in the rocket-based space launch industry is mainly performed through three contractual practices designed to handle and minimize the numerous risks inherent to this space operation and to protect each party involved in it, from the major parties in the launch agreement to their respective subcontractors. These practices are the concept of best efforts, the cross-waiver of liability, and the hold-harmless agreement. These three contractual agreements have constituted a common basis for all launch agreements since the 80's, and is directly inspired by the Launch Services Agreement for NASA's Space Transportation System [6]. Will these practices be required to handle legal risks for the Space Elevator system?

3.1.   The Principle of Best Efforts

The "best efforts" clause, by which the launch operator commits himself to perform the operation with all the means at its disposal, furnishing its best efforts, introduces two main principles that are inherent to the space launch industry. First, it emphasizes the need for cooperation between contractors. Second, it acknowledges the high level of risk inherent to the launch activity. By utilizing this clause, the launch operator is not bound by a certain result, but only by the way it has to perform its contractual obligations. Thus, the company has to follow a strict duty of care, which is realistic and acceptable considering that the operator has no control over the operation as soon as the rocket begins flight.

Some authors [7] view this clause as an exoneration of liability; the operator being discharged of all obligations in the performance of the launch. However, being classified as an "open terms contract", which means that the content of the notion of best efforts is moving with the technological evolution in the sector [8], it is more likely that the concept has to be strictly interpreted. Furthermore, the general understanding of the best efforts requires from the parties that they act with the highest devotion and according the highest standards of quality while performing their contractual obligations [9]. An appropriate definition of the concept would thus be "diligently working in a good and workman-like manner, as a reasonable, prudent manufacturer of launch vehicles and provider of launch services" [10] but not bound by a specific result. These divergences in the interpretation of the notion explain why it does not expressly appear anymore in the launch agreements [11].

The concept of best efforts is vital for launch service providers, because they have little or no control on the launch process once it has started. Assuming the Space Elevator will enable the operator to control the entire process, being even able to stop or reverse it, is the notion of best efforts still relevant?

This question will have to be answered as the project becomes more viable. However, it seems possible that the Space Elevator will be reliable and controllable enough to remove the notion of best efforts from the operator's contractual obligations. Moreover, as customers become more and more reluctant to accept the notion of best efforts, by guaranteeing results in the launch agreement, customers would be incentivized to choose the Space Elevator system instead of a traditional rocket-based launch. The final question is then whether or not the Space Elevator operator would be able to obtain the necessary insurance coverage in the case that it would incur liability as the result of such obligations. Ultimately, the answer will depend on the technical reliability of the system.

3.2.   The cross-waiver of liability and the hold-harmless agreement

The cross-waiver of liability and the hold-harmless agreement are two other clauses included in every launch agreement as a legal risk management technique that effectively compartmentalizes the risks between the parties to the launch agreement and their respective subcontractors. According to the cross-waiver of liability, each party to the launch agreement will support all corporal or material, direct or indirect, damages suffered by itself or its partners and caused during the execution of the contract. By including this clause to the launch agreement, the parties expressly agree not to engage in litigation against the other party or its partners, for any reason whatsoever [12].

According to the hold-harmless agreement, in case one or more partners of a party bring litigation against the other party to the launch agreement or its own partners, the first party will guarantee the other or its partners for the outcome of the lawsuits, bring the funds necessary for the defense of its interests and cover the cost of any judgment.

These clauses to the contract compartmentalize the risks between the two contractual chains [13], meaning that each party will cover its own damages, protect its partners and avoid litigation in the case of a launch failure. This instrument of legal risk management in rocket-based launch services has proven efficient, by preserving a highly technical (even experimental) industry from the multiple risks it suffers, allowing the emergence of new actors, the development of space activities in general, and maintaining commercial viability in the operation of rocket-based launch systems even after multiple failures.

Various factors must be considered in order to determine whether or not the cross-waiver of liability and the hold-harmless agreement should be included in the Space Elevator agreement. The primary question is whether such protection is necessary, considering the risk profile of the Space Elevator system?

Assuming that the Space Elevator system will be more stable and create less vibration than a rocket-based launch system, the risk of damage caused by the launch vehicle should be greatly reduced, if not null [14], also reducing the risks for the satellite owner to be held liable for damages to the Space Elevator. The same conclusion applies to the Space Elevator operator, whose chances of completing with success [15] the placement of the payloads in an orbital position should greatly increase with a reliable Space Elevator rather than with a rocket-based launcher [16]. The cross-waiver of liability between the two main parties to the launch agreement may be discarded as increased reliability of the system removes the necessity of these contractual practices, against which many legal systems have always been reluctant in fields other than commercial space launch.

Moreover, removing these clauses from the agreement would not mean an increased cost of insurance coverage (once again assuming that the Space Elevator will be almost perfectly reliable) and would put an end to the controversy regarding their validity and legality [17]. It would constitute another incentive for satellites owners to choose the Space Elevator instead of a rocket-based launch system, which could be vital at the outset of the operation of the system. The Space Elevator could attract clients by demonstrating that the reliability of the system makes the traditional methods of risk management unnecessary.

However, an increasing number of space faring States include these practices in their legal instrument on space operations, even defining them as mandatory. The Space Elevator operators will have to review these regulations in case they effect the operation of the Space Elevator. Space faring nations may need to create specific legal instruments regarding liability for "launch" services provided by the Space Elevator.

Another obstacle to this exclusion from the launch agreements is the need of subcontractors for legal protection against possible product liability. Considering the vast amount of liability that could arise in case of the failure or destruction of the structure linked to a defective product furnished by a subcontractor, traditionally a weak party, finding companies willing to participate in the construction of the structure would be a challenge if subcontractors could be subjected to product liability. As it is required for rocket-based launch activity, the successful operation of the Space Elevator calls for protection of subcontractors.

In the end, questioning the legitimacy (and the usefulness) of the cross-waiver of liability between the two parties to the Space Elevator agreement may be justifiable depending of the ultimate characteristics of the structure. However, reallocating the risk among the parties would be a mistake if doing so jeopardized the development of the Space Elevator project.

4.   **Legal Framework on Space Operations**

States began to enact legal instruments early on to implement the obligations contained into international conventions [18] on space activities they had ratified, as well as to regulate the activities of private actors in order to ensure a maximum level of security in the accomplishment of these space operations. Norway was the first country to enact such legal instruments in 1969, followed by the United States in 1984 with the Commercial Space Launch Act [19], Russia in 1993, South Korea in 2007 and the latest being France with its law on space operations in 2008 [20].

Because the legal regime governing the Space Elevator cannot be known with certainty today [21], it is not possible to state with precision which legal obligations could be applied to it. However, most of these legal instruments share a common structure, due to the influence of the United States, such as: the registration of space objects, authorization of space operations by an administration and licensing procedures, surveillance of space operations performed by private entities by the relevant administration, and establishment of a regime of liabilities concerning space operations. This structure has proven in the past thirty years to be an effective instrument of risk management in the sector of space operations. It is likely that it will be adapted to the Space Elevator as necessary.

The registration of the Space Elevator on the national and international registers [22] should not require a change in the process. Nevertheless, attention should be given to the sharing of information concerning the Space Elevator's position at the international level. A rocket-based launch system doesn't need to be given a specific position in outer space because, once it has performed its mission and reenter into the atmosphere, it is destroyed.

The Space Elevator will be permanently in outer space, occupying a position that could interfere with the trajectories of space objects. It is more similar to a satellite than to a launcher. The risk of interfering with other space objects' trajectories will require cooperation at the international level. The International Telecommunication Union, which is in charge of managing orbital positions, would probably be the most appropriate entity to perform this activity. International cooperation will also be required for the surveillance of the area surrounding the Space Elevator in order to avoid any collision with in-orbit satellites and space debris.

The authorization and licensing process will be the most important part of the legislation governing the Space Elevator. Since the legal act in itself usually sets only general principles, the regulation containing the technical requirements needed to obtain the authorization and the license needed to operate a space object will determinate the entire authorization and licensing procedure. Two issues will have to be considered in the drafting of this regulation: safety and economic viability. The appropriate balance between these two concerns will have to be obtained in order to ensure that the Space Elevator can operate sustainably.

The establishment of a regime of liabilities is also an important risk management tool, since it introduces predictability into a hazardous activity such as commercial space launch. It must be determined whether or not the cross-waiver of liability, which is mandatory for commercial space launch agreements in many national laws on space operations, should be maintained as it stands today or minimized so that it only ensures the protection of the subcontractors of the main parties to the agreement.

Legal instruments on space operations usually impose on the operators an obligation of insurance coverage, which will likely apply to the Space Elevator as it is a major instrument of risk management by limiting the exposure to liability of the parties. However, will the space insurance market be able to cover the risks of damages to the structure, or even its destruction, considering the cost of building a Space Elevator and the theoretical limit of the market as it stands today [23]?

After a medium or even short operation of the Space Elevator, assuming that the rates of launches and the reliability of the system will be highly increased compared with rocket-based launchers, the insurance companies should be able to cover the risks of the destruction of the structure thanks to an increase in their profits from the space sector. However, will a private company be able to gather the necessary investment and operate such an innovative system without full insurance coverage of the risks created by it?

5.   Conclusion

As contemplated today, the Space Elevator project could lead to a revolution in space activity and human societies. But behind the scientific and sociological aspects lies an economic reality. To be developed, the Space Elevator must appear to be an economically viable and profitable project in order to gather the necessary investments and support from States, companies, and the existing space-launch industry. Every business involving a high level of technical and legal risk needs an effective management regime that reduces the exposure to those risks. To create such a regime, there should be a preliminary study of the existing industry. The structure of the Space Elevator legal risk management regime may be greatly influenced by the techniques used for the rocket-based launch industry, after adjustments as necessary or relevant.

It appears that the this new means of transportation would greatly reduce the current risk in placing payloads in outer space, which should influence the way we view the Space elevator. Discarding the unnecessary or antiquated means of risk management may represent an effective incentive for clients to choose the Space Elevator over traditional rocket-based launch systems. This may ultimately contribute to the commercial success of the project by lowering the costs of liability to the actors and proving the high reliability of the system.

References

[1] Romain Loubeyre is a Law student from the University Paris-Sud 11 specialized in International and European Law as well as in Space and Telecommunication Law. He is a member of the Institute of Space and Telecommunication Law from the Law Faculty Jean Monnet since 2010. His researches include subjects as the regime of legal risk management in commercial space launch activity, the space insurance market or the influence of the French law n°518-2008 (April 3rd 2008) on the space operations industry.

[2] For a presentation of risk management in the space industry: S. GEROSA et F. NASINI, « Project Financing and Risk Management: a new challenge for program management in the space industry of the third millennium », PM World Today, Vol XI, n°8, second edition, August 2009, p. 1.

[3] Especially if the anchor station is located in the high seas, in this way: B. Jarrell, International and Domestic Legal Issues Facing Space Elevator Deployment and Operation, 7 Loy. L. & Tech. Ann. 71 (2007), p.87.

[4] For some technical issues on the Space Elevator: Dr J. LoSecco, Technical Issues Basic Space Elevator Physics, in Liftport: The Space Elevator--Opening Space to Everyone, by Michael J. Laine, B. Fawcett, T. Nugent, LiftPort Media, Inc., 2006.

[5] The usual amount referred to is around US$ 10,000,000,000

[6] Droit des activités spatiales, adaptation aux phénomènes de commercialisation et de privatisation, directed by L. RAVILLON, Université de Bourgogne – CNRS, Study of the Center of research on markets regulation and international investments, Volume 22, Edition LexisNexis Litec, 2004, p.494.

[7] J. O'BRIEN, « The structuring of NASA Launch Contracts », in Aspects juridiques et commerciaux des activités spatiales, Institut du droit et des pratiques du commerce international et International Bar Association, meeting of Paris, December 5th and 6th 1985, Paris, ICC, 1986, p.5.

[8] Queen's bench division, Midland Reclamation Limited c. Warren Energy Limited, 20 Janvier 1997, Official Referees' Business n°254.

[9] B. SCHMIDT-TEDD, « Best Efforts Principle and terms of contract in Space Business », International Institute of Space Law journal, n°31, 1998, p.330.

[10] T.L. MASSON, « The Martin Marietta Case or How to safeguard Private Commercial Space Activities », International Institute of Space Law journal, n°35, 1992, p.247.

[11] Which does not mean that the operator is not subject to the best efforts principle anymore. On the contrary, excluding the clause from the launch agreement simply allowed to avoid any controversy over the interpretation of the notion.

[12] Y. AUBIN and T. PORTWOOD, « les clauses réciproques d'abandon de recours et de garantie contre les recours des tiers », Revue de Droit des affaires internationales, n°6, Thomson/Sweet & Maxwell, 2001, p. 671.

[13] One of the characteristics of the commercial space launch activity is the presence of two contractual chains: the first one being the launch operator and its partners/subcontractors; the

second one being the satellite owner and its partners/subcontractors. The two chains are linked by the launch agreement concluded between the launch operator and the satellite owner.

[14] Many launch failures due to the satellite were due to high vibrations of the payload damaging the rocket-based launcher.

[15] By completing with success, we mean the placement of a satellite that exactly fits the specificities contained in the launch agreement (exact orbital position, Stated Satellite Life (SSL), technical capabilities fully preserved and operational...).

[16] Even though most of the current launchers used by experienced operators have a success rate that can reach more than 90%, the risks of a partial loss of the satellite's capabilities due to damages experienced during the launch process are still significant.

[17] P. DELEBECQUE, « les renonciations à recours », in Etudes offertes au Doyen Philippe Simler, Paris, LexisNexis, Dalloz, 2006, p.571.

[18] On the international treaties regarding the use of outer space applied to the Space Elevator: B. Jarrell, International and Domestic Legal Issues Facing Space Elevator Deployment and Operation, 7 Loy. L. & Tech. Ann. 71 (2007) p. 90.

[19] Commercial Space Launch Act, in the United States Code, Title 49, Chapter 701, Section 70101, as amended on December 23nd 2004.

[20] Law n° 2008-518 of June 3nd 2008 on space operations, published in OJ n° 129 of June 4th 2008.

[21] On the US legal system applied to the Space Elevator, see: B. Jarrell, International and Domestic Legal Issues Facing Space Elevator Deployment and Operation, 7 Loy. L. & Tech. Ann. 71 (2007) pp. 75-87.

[22] The international register of space objects is held by the Secretary General of the United Nations.

[23] The evaluation was around US$ 785,000,000 on March 2010 (interview with C. Wells, Legal & Claims Manager at SCOR Global P&C).

MODELING THE SELF-HEALING OF BIOLOGICAL OR BIO-INSPIRED NANOMATERIALS

Nicola M. Pugno[1,2,3,*] and Tamer Abdalrahman[1]

[1]Laboratory of Bio-Inspired Nanomechanics "Giuseppe Maria Pugno", Department of Structural Engineering and Geotechnics, Politecnico di Torino, Corso Duca degli Abruzzi 24, 10129, Torino, Italy.
[2]National Institute of Nuclear Physics, National Laboratories of Frascati, Via E. Fermi 40, 00044, Frascati, Italy.
[3]National Institute of Metrological Research, Strada delle Cacce 91, I-10135, Torino, Italy.
[*]nicola.pugno@polito.it

Abstract: Self-healing materials are a class of solids that have the capability to repair damage autonomically, as often observed in living materials. Here, a first model is presented that incorporates self repairing fibers to determine the expected mechanical behavior of a self-healing bundle.

1.   Introduction

Biological systems have the ability to sense, react, regulate, grow, regenerate, and heal. Recent advances in materials chemistry and micro- and nanoscale fabrication techniques have enabled biologically inspired materials systems that mimic many of these remarkable functions. Self-healing materials are motivated by biological systems in which damage triggers a site-specific, autonomic healing response. Self-healing has been achieved using several different approaches for storing and triggering healing functionality in polymers. There are different models for the prediction of the fatigue behavior of self-healing polymers [1-3].

Other classes of synthetic materials can undergo healing processes, which in mechanics are basically the mechanisms leading to the recovery of strength and stiffness after damage. However, most synthetic materials require outside intervention such as the application of heat or pressure to initiate and sustain the healing process. For example, Ando et al. [4-7] have shown the healing capability of sintered ceramics while exposed to high temperatures (1000 °C).

In addition, supramolecular materials naturally feature so-called «reversible» (non-permanent) intermolecular bonds, in contrast with polymers derived from traditional chemistry, which are based on so-called «irreversible» (permanent) bonds. This reversibility feature imparts a natural capacity to self-heal: cracks or breaks occurring in supramolecular materials can be repaired simply by putting the fractured surfaces back together and applying light pressure; the material nearly recovers its initial strength without the need for bonding or heating.

Too model in general self-healing materials, fiber bundle models can be used. A large number of non-healing models exist for fiber bundles [e.g. 8-11]. In contrast, according to the authors knowledge, there is no model for the prediction of the tensile behavior of self-healing fiber bundles.

This model is the aim of the present letter.

2.   Engineering self healing parameter

For a large number, N_0, of fibers in a bundle, the number of surviving fibers N_{s0}, under an applied strain ε, is given by [2]:

$$N_{s0} = N_0 \exp[-(\frac{\varepsilon}{\varepsilon_0})^m]$$ (1)

where ε_0 and m are the scale and shape parameters of the Weibull flaw distribution.

The fraction of broken fibers is given by:

$$N_{s0} = N_0 \exp[-(\frac{\varepsilon}{\varepsilon_0})^m]$$ (2)

and in case of self-healing:

$$\alpha_h = \frac{N_{bh}}{N_0} = \frac{N_0 - N_{sh}}{N_0}$$ (3)

where N_{sh} is the actual number of surviving fibers in the presence of self-healing.

Note that eqs. (2) and (3) resemble the definition of an engineering strain ($\varepsilon = \frac{l - l_0}{l_0}$).

We introduce the parameter λ, as the ratio between the number of broken fibers with self-healing, N_{bh}, and the number of broken fibers without healing N_{b0}:

$$\lambda = \frac{\alpha_h}{\alpha_0} = \frac{N_{bh}}{N_{b0}} = \frac{N_0 - N_{sh}}{N_0 - N_{s0}}$$ (4)

Finally, we introduce the healing parameter η, as:

$$\eta = 1 - \lambda = \frac{N_{sh} - N_0}{N_0 - N_{s0}}$$ (5)

Note that when $\eta=1$ we have $N_{sh}=N_0$, whereas for $\eta=0$, $N_{sh}=N_{s0}$.

3.   True self healing parameter

We now introduced the true parameter α_h^* as:

$$\alpha_h^* = \int_{N_0}^{N_{sh}} \frac{dN}{N} = \ln N_{sh} - \ln N_0 = \ln \frac{N_{sh}}{N_0} \tag{6}$$

in analogy with the true strain ($\varepsilon = \int_{l_0}^{l} \frac{dl}{l} = \ln \frac{l}{l_0}$).

In absence of healing it becomes:

$$\alpha_0^* = \ln \frac{N_{s0}}{N_0} \tag{7}$$

From equations (4) and (5), the true self-healing parameter is given by:

$$\eta = 1 - \lambda = 1 - \frac{\alpha_{sh}^*}{\alpha_0^*} = 1 - \frac{\ln \frac{N_{sh}}{N_0}}{\ln \frac{N_s}{N_0}} \tag{8}$$

The introduction of the true self-healing parameter of eq. (8) is needed in order to take into account the variation of the total number of fibers induced by the self-healing (similarly to the true strain that is accounting for the length variation).

From equation (1) we immediately derive:

$$\ln \frac{N_s}{N_0} = (-(\frac{\varepsilon}{\varepsilon_0})^m) \tag{9}$$

By substituting equation (9) into equation (8) we find:

$$\ln \frac{N_{sh}}{N_0} = ((\eta - 1)(\frac{\varepsilon}{\varepsilon_0})^m) \tag{10}$$

and thus:

$$N_{sh} = N_0 \exp[(\eta - 1)(\frac{\varepsilon}{\varepsilon_0})^m] \tag{11}$$

The introduction of the self-healing into eq. (11) generalizes the classical Weibull approach [12], eq. (1).

The last expression is related to the applied tensile load, P, by:

$$P(\varepsilon) = AE \; \varepsilon[N_0 \exp[(\eta - 1)(\frac{\varepsilon}{\varepsilon_0})^m]] \tag{12}$$

where A is the cross sectional area of the single fiber and E is its Young's modulus. Then, if A, L, E, N_0, m and ε_0 are known, the curve stress vs. strain can be obtained:

$$\sigma(\varepsilon) = \frac{P(\varepsilon)}{AN_0} = E \; \varepsilon[\exp[(\eta - 1)(\frac{\varepsilon}{\varepsilon_0})^m] = E_{eq}(\varepsilon, \eta)\varepsilon \tag{13}$$

4. Results and discussion

As an example we apply our calculation to carbon nanotube (CNT) bundle with strength randomly assigned, $\varepsilon_0 = 0.04$ and $m_0 \approx 2.7$, based on the nanoscale Weibull distribution [13].

Fig.1 shows the mechanism of the self-healing of a carbon nanotube. Self-healing of CNTs may accelerate the development of the CNT apace-elevator mega cable [15-17].

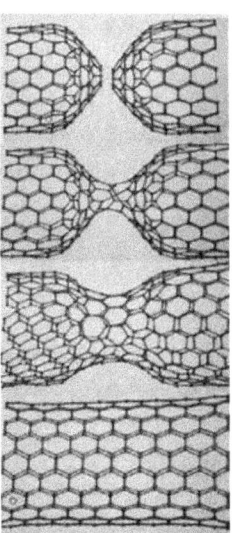

Fig. 1 Carbon nanotube self-healing mechanism [14].

In Fig. 2, the stress–strain response is predicted for a bundle with different values of the healing parameter, η, from 0 to 1, while all the other parameters in equation (13) are kept constant. When increasing the self-healing parameter, both the maximum stress, see Fig. 2, and the strain at which the maximum stress is reached, increase. (This can also be seen in Fig. 7, where the ratio between the maximum stress with healing and maximum stress without healing is increasing in a monotonic way with an increase in the healing parameter.) For a self-healing parameter equal to 1 the bundle becomes unbreakable.

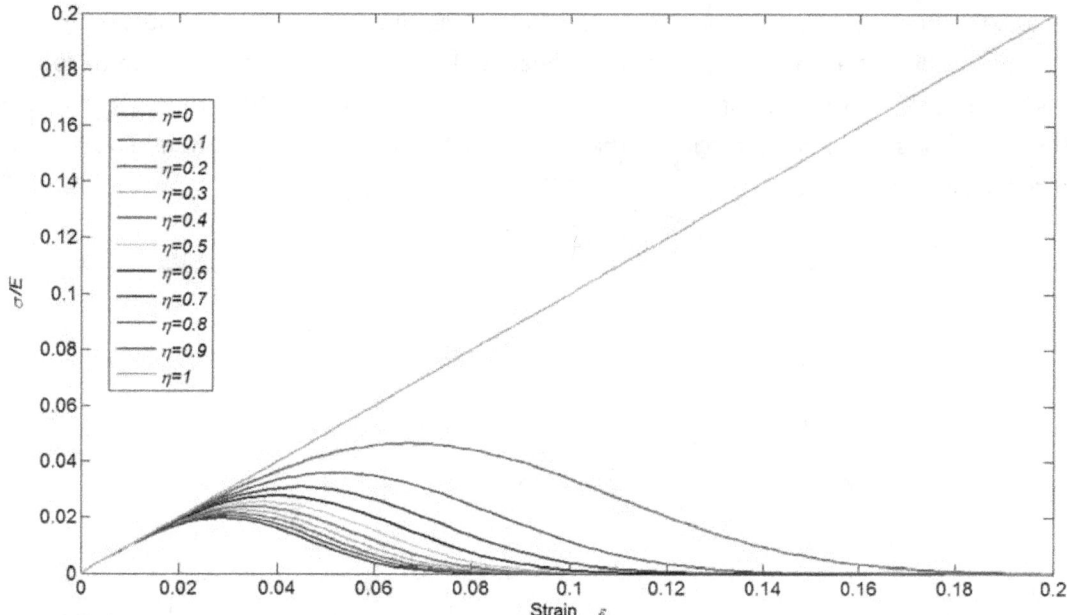

Fig. 2 Stress-strain response by varying the self-healing parameter.

Fig. 3 shows the variation of the number of survival fibers as a function of the applied strain, with different values of the healing parameter.

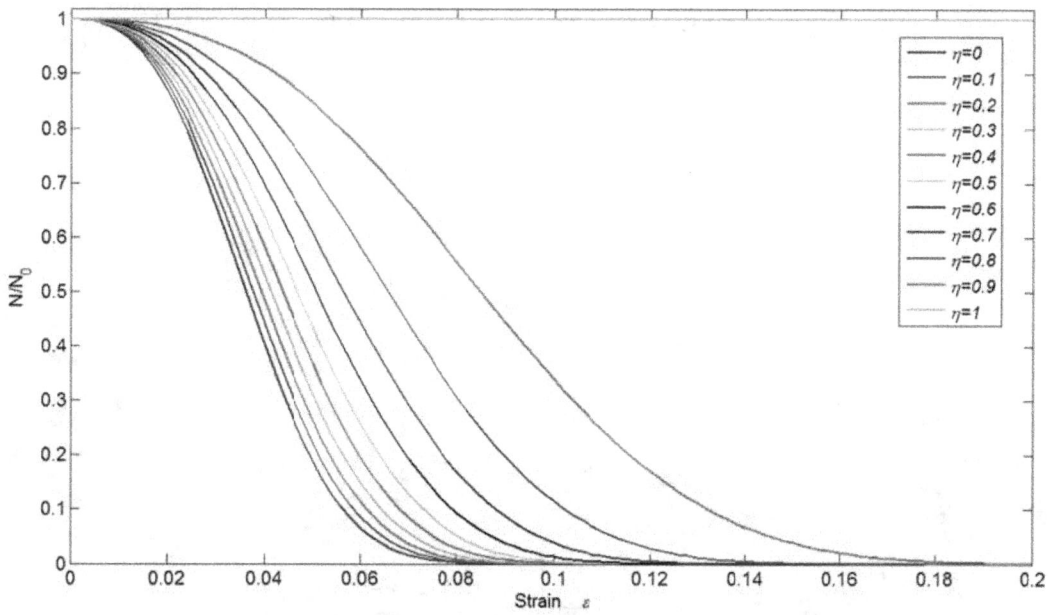

Fig. 3 Survival fibers, $N_{sh} \equiv N$ vs. strain, by varying the healing parameter.

Fig. 4 shows two different type of curves. The upper curves represent the stress-strain curves of Fig. 2 and the lower curves are the rates of variation of the number of survival fibers in the bundle, by varying the applied strain and for different self-healing. The maxima of the lower curves represent the points at maximal failure rate of the bundle. From Fig. 4 we can see that the strains at which the maximum stress is reached, ε_σ^*, are lower than the strains at the maximal failure rate, ε_n^*, as specifically reported in Fig. 5.

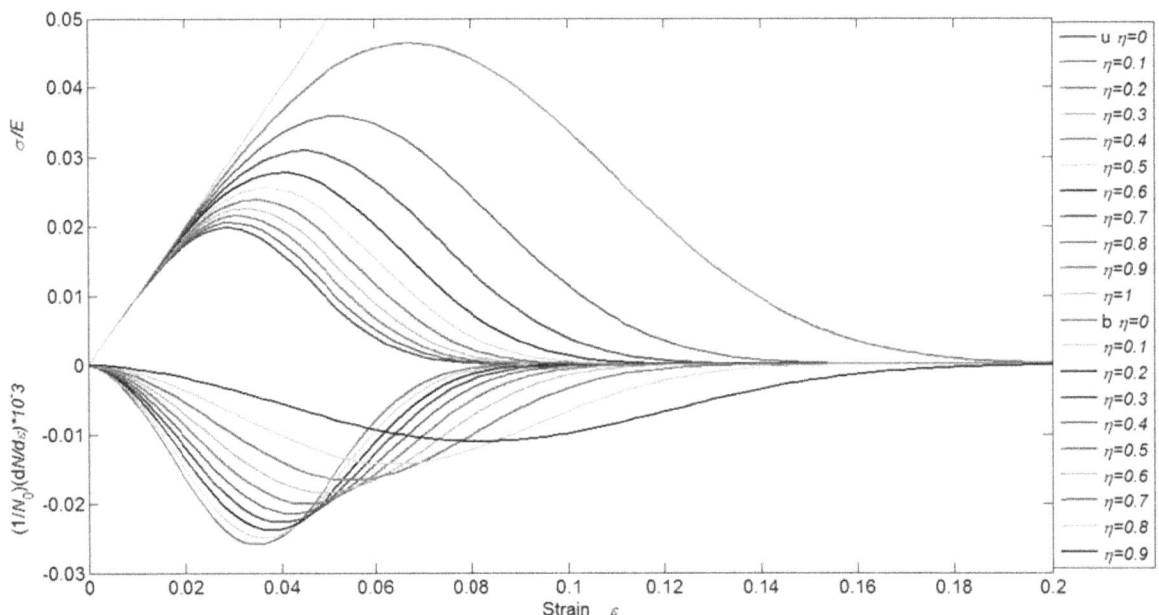

Fig. 4 Stress or rate of survival fibers vs. strain by varying the healing parameter.

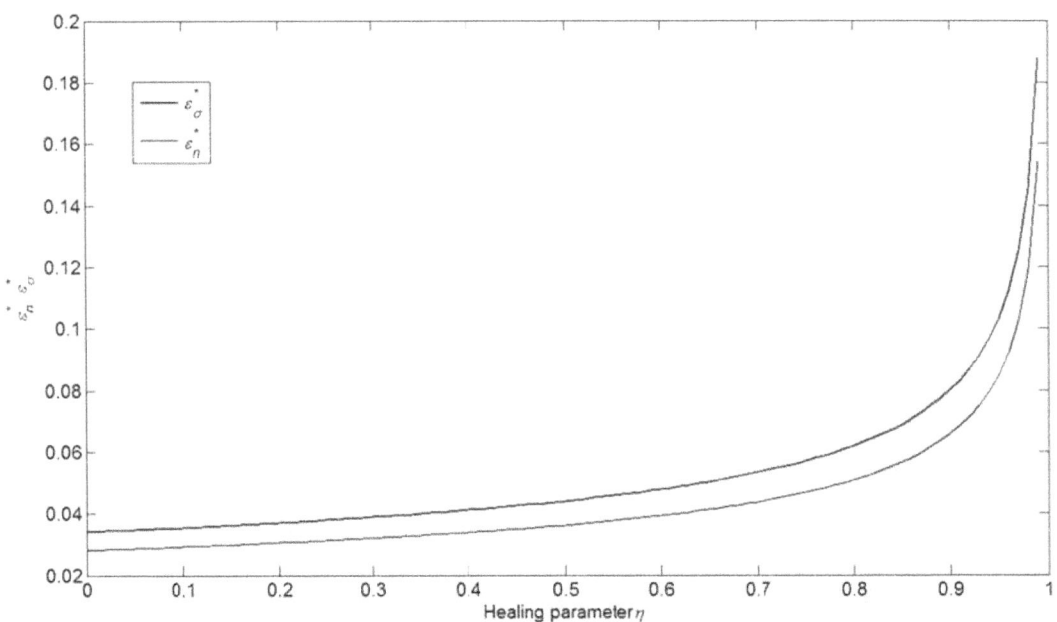

Fig. 5 Strains corresponding to maximum stress or failure rate vs. healing parameter.

The area under the stress-strain curve is the total dissipated energy density (in our calculations we assumed that the bundle is fractured when the stress is 1% of its maximum). In Fig. 6 the ratio between the dissipated energy density with and without healing is reported and clearly increases with increasing the self-healing parameter.

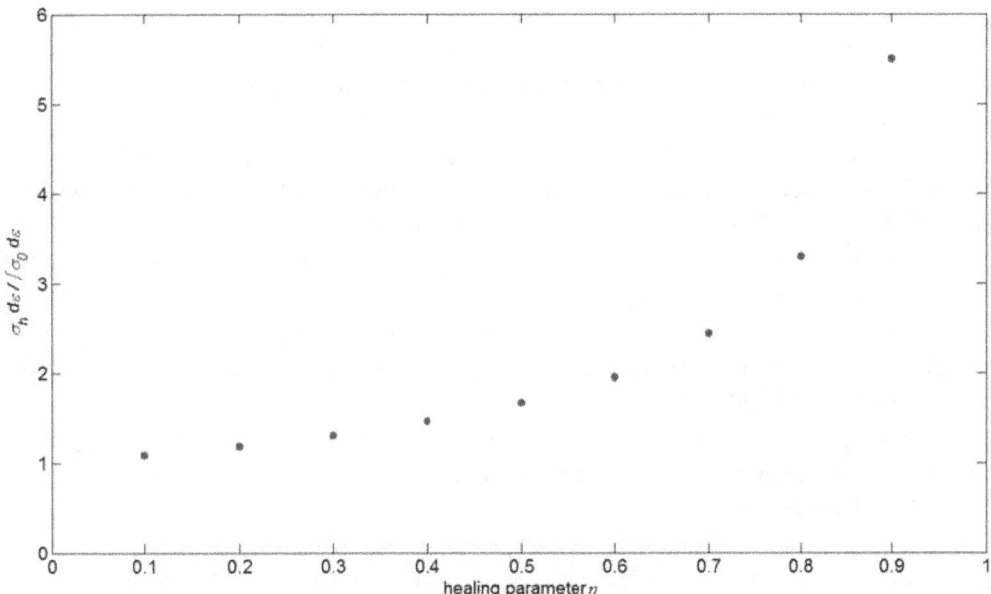

Fig. 6 Dissipated energy density with self-healing normalized to the non healing case vs. healing parameter.

The ratio between the strains corresponding to the maximum stresses with and without healing $\varepsilon_{max,h}$ and $\varepsilon_{max,0}$ respectively and the ratio between the related maximum stresses $\sigma_{max,h}$ and $\sigma_{max,0}$ respectively are reported in Fig. 7: both these ratios increase by increasing the healing parameter.

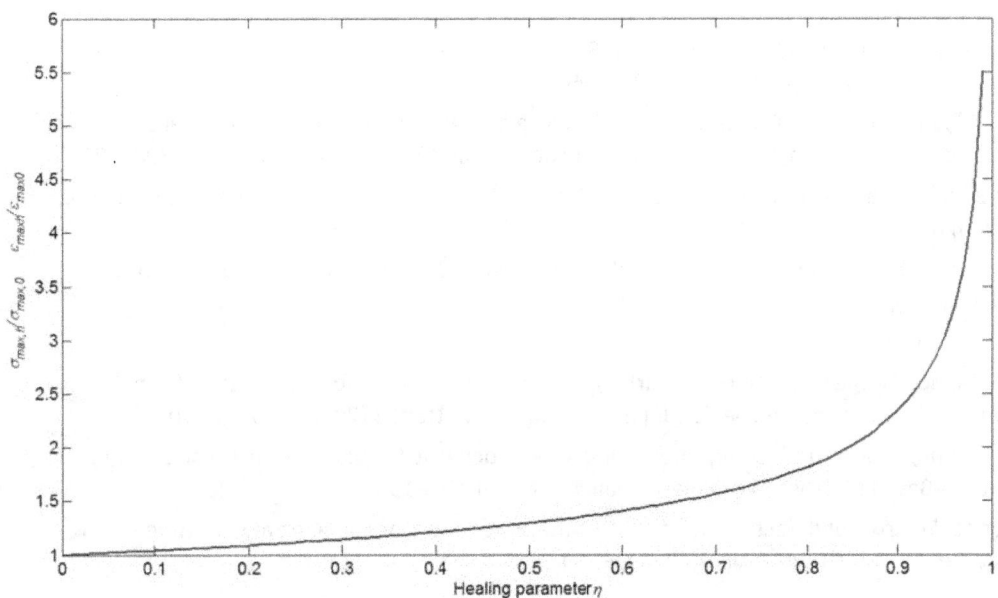

Fig. 7 Dimensionless maximum stress and related strain vs. healing parameter.

5. Conclusions

The presented simple self-healing fiber bundle model is able to quantify the increments of the mechanical performances induced by the self-healing. Applications to design a new class of bio-inspired nanomaterials are envisioned.

References

[1] S. Maiti, P. Geubelle. A cohesive model for fatigue failure of polymers. Engineering Fracture Mechanics 72 (2005), 691–708.

[2] S. Koussios, A. J. M. Schmets. Hollow helix healing: A novel approach towards damage healing in fiber reinforced materials. Proceedings of the First International Conference on Self Healing Materials 18-20 April 2007, Noordwijk aan Zee, The Netherlands.

[3] A.S. Jones, H. Dutta. Fatigue life modeling of self-healing polymer systems. Mechanics of Materials 42 (2010), 481– 490.

[4] K. Ando, M. Chu, T. Hanagata, K. Tuji, S. Sato. Crack-healing behavior under stress of mullite/silicon carbide ceramics and the resultant fatigue strength. Journal of the American Ceramic Society 84 (2001), 2073–2078.

[5] K. Ando, M. Chu, K. Tsuji, T. Hirasawa, Y. Kobayashi, S. Sato. Crack healing behaviour and high-temperature strength of mullite/SiC composite ceramics. Journal of the European Ceramic Society 22 (2002), 1313–1319.

[6] K. Ando, M. Chu, K. Tsuji, T. Hirasawa, Y. Kobayashi, S. Sato. Crack healing behavior of Si3N4/SiC ceramics under stress and fatigue strength at the temperature of healing (1000 °C). Journal of the European Ceramic Society 22 (2002), 1339–1346.

[7] K. Ando, M. Chu, K. Tsuji, T. Hirasawa, Y. Kobayashi, S. Sato. Crack healing behavior of Si3N4/SiC ceramics under cyclic stress and resultant fatigue strength at the healing temperature. Journal of the American Ceramic Society 85 (2002), 2268–2272.

[8] Z. Chi, T.W. Chou, G. Shen. Determination of single fiber strength distribution from fiber bundle testings. Journal of Materials Science 19 (1984), 3319–3324.

[9] T. Xiao, Y. Ren, K. Liao, P. Wu, F. Li, H.M. Cheng. Determination of tensile strength distribution of nanotubes from testing of nanotube bundles. Composites Science and Technology 68 (2008), 2937–2942.

[10] A. Cowking, A. Atto, A. Siddiqui, M. Sweet, R. Hill. Testing E-glass fiber bundles using acoustic emission. Journal of Materials Science 26 (1991), 1301–1310.

[11] M.R. Mili, M. Moevus, N. Godin. Statistical fracture of E-glass fiber using a bundle tensile test and acoustic emmision monitoring. Composite Science and Technology 68 (2008), 1800–1808.

[12] W. Weibull. A statistical theory of strength of materials. Royal Swedish Institute of Engineering Research (Ingenioersvetenskaps Akad. Handl.), Stockholm, 153 (1939), 1–55.

[13] N. M. Pugno and R. S. Ruoff. Nanoscale Weibull statistics. J. Appl. Phys. 99 (2006), 024301– 4.

[14] http://eurospaceward.org/2010/Summary_report_on_4th_ESW_conference.pdf. Prof. B. Yakobson's talk.

[15] N. M. Pugno, On the strength of the carbon nanotube-based space elevator cable: from nanomechanics to megamechanics, J. Phys.: Condens. Matter 18 (2006), 1971–1990.

[16] N. M. Pugno, F. Bosia and A. Carpinteri. Multiscale Stochastic Simulations for Tensile Testing of Nanotube-Based Macroscopic Cables. Small 4 (2008), 1044 –1052.

[17] N. Pugno. The role of defects in the design of the space elevator cable: from nanotube to megatube. ACTA MATERIALIA (2007), 55, 5269-5279.

Ben Shelef, the Spaceward Foundation

Abstract: This paper ties together parameters pertaining to tether specific strength and to power system mass density to arrive at an inequality that determines whether a space elevator system is viable.

The principle for the feasibility condition (FC) is that a space elevator must be able to lift its own weight fast enough – fast enough to grow by bootstrapping, fast enough to replace aging material, and fast enough to have a significant margin for commercial cargo beyond these housekeeping tasks. The FC therefore sets a 3 dimensional design space comprised of {tether material specific strength, power system specific power, system time constant}.

After developing the FC, real life limitations on specific power and specific strength are plugged in, and the resultant viable design space is examined. Finally, a design architecture that satisfies the Feasibility Condition is briefly introduced.

1.   Motivation

As is well known, there is no hard minimum requirement on the specific strength of a space elevator tether. The lower the specific strength, the higher the tether taper ratio, and the heavier the tether gets for a given climber mass. Previous work (Edwards, [4]) cites a UTS of 130 GPa for CNTs, and a density of 1.3 g/cc, for a specific strength of 100 GPa-cc/g, and a taper ratio of 1:2. Edwards uses this number as a starting working point, with the implication that if CNT tethers will not reach this specific strength, the effect will be an increase in the taper ratio and in the total mass of the space elevator tether, but this will not be a fundamental problem or show-stopper for the construction of the space elevator.

In a similar fashion, there is no absolute requirement on the performance of the power system of the space elevator. Higher specific power allow the climber to move faster and clear the bottom of the tether sooner, increasing the launch rate and mass throughput of the system. In [4] Edwards cites a 2 MWatt power system weighing 5 tons (0.4 kWatt/kg), able to launch a climber about once a week. Again, the implication is that if this specific power cannot be reached, the only penalty will be a reduction in the possible payload mass throughput, but this will not affect the feasibility of the space elevator.

In previous discussions of space elevator design, tether strength and power systems are treated as mostly independent domains – railroad tracks and train engines, to use a familiar analogy.

There is, however, another assumption in the space elevator architecture that ties these two domain together. It is accepted that a space elevator is too heavy to launch directly, and so the only way to construct a viable-sized space elevator is to launch a smaller seed space elevator, and use its lifting capacity to bootstrap to a much larger space elevator. In addition, the tether material will have a certain expected lifetime in operation, and so the entire tether mass has to be replaced at a certain minimum rate, using the space elevator itself to perform the task.

These "housekeeping" lift chores create the link between the tether system and the power system, since a lower-performance tether requires more housekeeping work, and thus levies a throughput requirement on the power system, thereby denying us the option to simultaneously have both an arbitrarily weak tether and an arbitrarily weak power system. If the housekeeping chores cannot be kept, then not just the performance, but indeed the feasibility of the space elevator is cast into question.

2.   Feasibility Condition

To lay out the rationale for the FC, we need to define some basic design parameters for the space elevator.

Since the space elevator is linearly scalable, the discussion below is independent of the size of the system. We therefore normalize all mass parameters by the maximum mass that is allowed to hang from the bottom of the tether (m_{max}) and refer to this unit as a *Standard Mass Unit* (SMU). Thus for example a 20-ton space elevator is one that can support a 20 ton load at ground level, and if its tether weighs 6000 tons, then we say the tether weighs 300 SMUs. The ratio of tether mass to lifting capacity is labeled *Tether Mass Ratio* (TMR) – 300 in this case.

The *Tether Specific Loading* (TSL) is similar to the tether material specific strength, but takes into account parasitic mass (such as cross-weaves) and the margin of safety, so is the effective tether specific strength used in the design. For example, the ribbon material may achieve 40 MYuri, but the TSL used in the design is only 30 MYuri. Given a specific TSL, and using the constant-stress space elevator tether formula, it is possible to calculate the taper ratio, total mass, and thus the TMR.

Shifting the attention to the power system side of the design, the *Payload Mass Throughput* (PMT) of the space elevator is defined as the amount of mass it can move per unit time. The normalized unit for PMT is the *Standard Throughput Unit* (STU), defined as one SMU per year. The PMT is highly dependent on the specific power (SP) of the power system, measured in kWatt/kg, but this dependency is more complex than the one between the TSL and TSR.

Simplistically, the mass of the climber is divided into payload and power system (neglecting structure and other mass overhead), and the *payload mass ratio* (PMR) is defined as the mass of the payload divided by the total climber mass. Optimal PMT depends on the PMR (a higher PMR means the climbers carry more payload, but move slower)

Optimal PMT also depends on the number of concurrent climbers on the tether. Since each climber further up the tether weighs much less than a climber that is just taking off, it pays to have multiple climbers, each being lighter than 1 SMU. (Thus for example a 20-ton space elevator can operate 5–6 concurrent 15 ton climbers). In this respect, optimal PMT is achieved by having a continuous stream of infinitesimal climbers, but the optimization process is capped much earlier by the interactions between the power system and the Earth's rotation, limiting climbers to a once-per-day schedule.

A more detailed treatment of PMT optimization is presented in [10]. We are using the case limited by a once-daily cycle, since further (unpublished) work in respect to direct-solar power systems is making this to be the likely case to be used.

It is thus possible to derive the TMR from the TSL (through the constant-stress formula) and the PMT from the SP (through optimization and some operational considerations). The FC will create a dependency between the TMR and the PMT.

The Space Elevator Feasibility Condition (FC) is built around the concept of the characteristic time constant (CTC) – the time it takes the system to lift a payload mass equal to the mass of its tether. With all of the above parameters defined, the CTC is simply the ratio of the tether mass and the payload mass throughput of the system: CTC = TMR/PMT. The reciprocal, 1/CTC, is therefore a measure of the "specific throughput" of a space elevator system. A space elevator with a CTC of 2 years can only lift half of itself into orbit each year.

The CTC can now be compared with several time constants that are required under some operational assumptions.

We call the periods where the space elevator bootstraps, either during initial construction or during recovery from a break, "growth periods", to be contrasted with "normal operations". (Bootstrapping is the process of growing the space elevator's tether using the existing tether as the transport mechanism)

During growth periods, the space elevator relies on a certain growth rate, or time-to-double (TD). It is generally accepted that the size of the rocket-launched seed elevator is about 5% - 10% of the size of the first viable space elevator. This means that the space elevator has to double in size 3–4 times between its seed state and its operational state. We would argue that for a commercial transportation infrastructure project, 8–12 years is a reasonable limit to the amount of time it can be under construction before carrying payloads, and so TD of 2–3 years is reasonable. Civil infrastructure projects (E.g. Golden Gate Bridge, Transcontinental Railroad, British Channel Tunnel) all took a similar number of (or fewer) years to complete.)

Since we want to hold a spare tether spool in orbit to enable reasonable recovery from a tether break event, we need to count on this additional mass being launched as well. We'll denote the fraction weight of the spare-in-orbit as FS. If, for example, FS = 25%, then it will take 2 doublings (and thus 2*TD years) to fully recover from the a broken tether.

There is no absolute requirement on TD and FS, but we'll proceed right now with the notional values TD=1.5 and FS=25%.

During normal operations, the tether material will have some degradation rate in space as a result of cosmic radiation, micro orbital debris, chemical interactions, thermal cycling, and simple mechanical wear and tear. These factors will result in an allowed material lifetime in service, denoted TL. If any specific portion of the tether is prone to a shorter lifespan, we'll amortize that into the overall TL figure according to the size of that segment. TL is not currently known and depends partially on specific tether properties, but it is reasonable to assume TL >> TD given the values above.

To support growth, the space elevator must lift (1+FS)/TD of its mass per year. In a similar manner, to replace aging material, the space elevator must lift 1/TL of its mass per year. These are the housekeeping lift requirements introduced above.

Since during growth periods the housekeeping lift requirements are much higher than during periods of normal operations, it is enough to require that the space elevator will exactly support the housekeeping operations during growth periods, and use the additional capacity during normal operations for lifting payload. The sum (1+FS)/TD + 1/TL is therefore the minimal *required* lift capacity of the space elevator, in units of ribbon-masses-per-year.

The Feasibility Condition can thus be phrased as follows:

$$\frac{PMT}{TMR} = \frac{1}{CTC_{pos}} > \frac{1}{CTC_{req}} = \frac{1+FS}{TD} + \frac{1}{TL}$$

Obviously, if the feasibility condition can be met, and since during normal operations only the second term (1/TL) is present, the throughput represented by the first term (TMR·(1+FS)/TD) is all available for payload transportation.

3.   CNTs and Power Beaming System Performance

In this section we'll try to estimate the values of PMT and TMR achievable in the foreseeable future. As such estimates go, these will only establish a very rough range, since the technologies are complex and we are trying to look rather far into the future. The goal here is to merely estimate these values to an order of magnitude, so we can determine whether the FC shows the space elevator as "easily feasible", "clearly infeasible", or somewhere in between.

3.1   Tether material

Based on a gradual convergence of experimental and theoretical results, the specific strength of raw CNTs [7] will not exceed 50 MYuri [8],[9] as compared to previous estimates of 100 MYuri[4]. In particular, a failure mechanism known as the Stone-Wales defect causes spontaneous defects in the Nanotube structure and limits the possible strength.

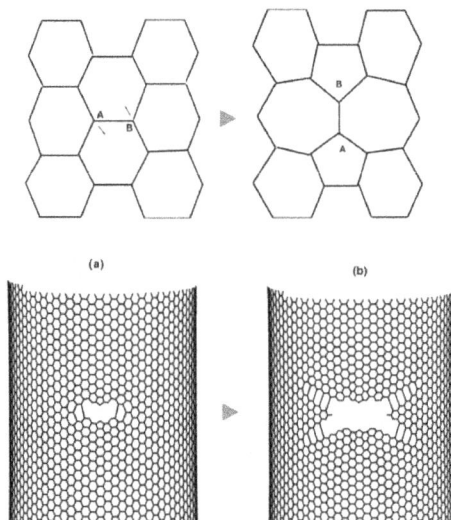

Figure 1: Stone-Wales Defect Formation

Using 45-50 MYuri CNTs, we can expect a near-flawless spun tether (A task that has not been achieved yet) to perform at 40 MYuri, and with a 33% safety margin, we can load the tether at a TSL of 30 MYuri. The weight of various redundancy structures can be shown to be only a few percent of the total tether mass, so will not affect this result significantly.

3.2   Power systems

The specific power of the power system is a function of the photovoltaic receiver, the electric motors, the power electronics, and any required heat-rejection systems. If one of these components has a significantly low specific power, it will be the heaviest component of the system and become the limiting factor.

Figure 2 shows a 400 m² thin-film PV receiver demonstrated by DLR. The panel weighs only 32 kg, including the booms and deployment mechanism. (It stows in the little suitcase at the center – note the people in the back for scale). The panel can provide 50 kWatt of electric power under A0 illumination, or 1.5 kWatt/kg. This type of thin-film panel currently converts sunlight at slightly under 10%. When used with monochromatic light, and at more than one sun intensity, performance will of course go up. Looking forward, it is conceivable that PV technology can eventually deliver 5 kWatt/kg under these conditions. It is important to note that the mechanical structure shown was designed for zero g, and a space elevator receiver will have a higher structural mass.

Figure 2: Thin Film for-space PV Panel

The best electric motors today achieve a specific power of just under 1.0 kWatt/kg, though there is no theoretical barrier preventing this number from improving. Specifically, CNT-based conductors can provide a large leap in performance in that respect, replacing the use of copper in motor windings. Even today's CNT conductors rival copper in specific conductivity. The lightest electrical motors, however, are not the most efficient ones, and since heat rejection carries its own mass penalty, a balance must be struck between specific power and efficiency.

Heat is generated in the system in two primary locations. On the PV receiver itself, and in the drive system. Heat generated at the receiver is already spread out over a large surface, and is radiated from where it is generated. Therefore the efficiency of the receiver and its allowed operating temperature jointly determine how much laser power can be placed on it.

Heat generated at the engine and power electronics has to be conducted from them to radiators. Looking at limiting values, the rejection heat area density (at 100% emissivity) is 0.7 kWatt/m^2 per side for a radiator operating at 330 K (60 C), and 1.0 kWatt/m^2 if operating at 360 K. There is no inherent requirement as to how thick radiator fins have to be, and so in principle the radiator can weigh less than 1 kg/m^2.

In practice, the mass of radiator systems is dictated by fluid-based heat distribution plumbing, and by the specific heat conductivity of the fin material. Today's state of the art power rejection systems are unacceptably heavy and inefficient due to the use of heavy ammonia cooling loops and Aluminum heat conductors. The heat rejection system of the ISS, for example, operates at less than 0.01 kWatt/kg. If the power system is 95% efficiency, than this value is equivalent to 0.2 kWatt/kg when considering drive power.

Since the specific heat conductivity of CNTs is so much higher than that of aluminum (better than 100x) it is clear that great strides can be made in this field relatively quickly when the need arises.

3.3   Required throughput

Table 1 shows the values of the required lift capacity (in normalized STU units) as a function of TD and TL as discussed above, keeping the assumption of FS=0.25. While TL and TD are not known, the possible range of required values for 1/CTC is rather narrow – between 0.5 and 1.5.

Table 1:

Required throughput values

$$1/CTC = (1+FS)/TD + 1/TL$$

FS=25%		TL		
		4	6	10
T D	1	1.50	1.42	1.35
	1.5	1.08	1.00	0.93
	2	0.88	0.79	0.73
	3	0.67	0.58	0.52

3.4   Predictions

Based on the above, we can comfortably bracket tether specific strength at 25 – 30 MYuri, resulting in a TMR of 228 to 144.

The analysis of the power system is less conclusive. PV receivers and electric motors deliver 1 kWatt/kg even today and keep getting better. CNT electrical conductors stand to reduce the weight of motors. Improvement in efficiency will reduce the requirements on the heat rejection system. The heat rejection system itself does not have a fundamental limit preventing it from reaching similar levels, but is currently far from 1 kWatt/kg, and will require the use of either CNTs or other lightweight highly-conductive materials. Looking forward 10-20 years, it is possible to have the entire system deliver 1–1.5 kWatt/kg.

4.   Parameter values

We can now take the estimates of section 3 and see how they fit into the FC.

Table 1:
Tether-driven vs. Power-driven constraints and their effect on throughput:

$$1/CTC = PMT/TMR$$

For example:
A space elevator constructed with a 30 MYuri tether and 3.5 kWatt/kg motors, can lift its own mass 1.7 times a year.

	TMR	A/A$_0$	TSL						
	Tether mass ratio	Taper ratio	Tether specific loading MYuri[1]						
optimistic	50	2.6	50	2.0	2.7	3.4	4.0	4.6	5.0
	77	3.4	40	1.3	1.8	2.2	2.6	3.0	3.2
CNT	144	5.0	30	.69	.94	1.2	1.4	1.6	1.7
	228	7.0	25	.44	.59	.75	.88	1.0	1.1
	433	11.3	20	.23	.31	.39	.46	.53	.58
pessimistic	739	17.3	17	.14	.18	.23	.27	.31	.34

0.5	0.7	1.0	1.5	2.5	3.5	kWatt/kg	Specific power	SP
100	135	170	200	230	250	STU	Optimized payload throughput	PMT
pessimistic		Thin-film receiver + motors			optimistic			

Light shaded values are optimistic, dark are pessimistic, values in the central boxes are probable.

As expected, strong tethers match weaker power systems and vice versa.

Given realistic technology values (TSL=25,30 MYuri and SP=1-1.5 kWatt/kg) we can see that 1/CTC ends up in the 0.75 – 1.4 range, and the CTC is in the range of 1.3 – 0.71.

5.   Conclusion

The Space Elevator Feasibility Condition is a sufficient but not a necessary condition for the viability of the space elevator as a practical transport system.

However, it is "strongly sufficient" in that it is able to define and capture the principal requirements on the strength of the tether and the specific power of the power system. Without it, we don't really have a sufficient condition, since a space elevator can be built out of any tether material

---

[1] A Yuri is the SI derived unit for specific strength. 1 MYuri = 1 N/Tex = 1 GPa-cc/g

(since at worst, the tether will simply weigh more) and using any power system (since at worst, the climbers will simply move more slowly.)

The above analysis shows that the possible values of CTC matches the required values very tightly, under the assumptions made. This means the feasibility condition can be satisfied, but not without difficulty. The connection between the tether and power systems means that the power system, and specifically the heat rejection system, is as important a design challenge as the design of the CNT tether.

Additionally, the Feasibility Condition alerts us to other aspects of the design of the space elevator. For example, the handling of bi-directional traffic and the disposition of climbers after they have carried their cargo. If the climbers had to come down the same tether, then this affects the throughput of the system, and thus factors into the FC. If the climbers use a second tether as a "down" elevator, then this tether increases the mass of the system since it also has to be maintained. If the climbers are cast off after each use, then we need to bring into account the cost of disposable climbers, which is significant.

The last observation ties the Feasibility Condition to a future financial study of the space elevator. We often portray the space elevator as a "free" system. The tether is an invariant, the climbers reusable, and the cost of electricity to power the laser can quickly be calculated to be negligible (especially if solar light is used for most of the trip).

However, for a full financial model we need to factor in the cost of tether replacement, and perform financial trade-offs such as choosing between reducing throughput, using more tether material (as in a "down" tether), or disposing of the climbers. This will result in a "Space Elevator Financial Viability Condition", which will impose a tighter bound on its real feasibility. While the values of the technology parameters are still not known, the mathematical model can be constructed today.

References

[1] Artsutanov, Y., "Into the Cosmos by Electric Rocket", *Komsomolskaya Pravda,* 31 July 1960. (The contents are described in English by Lvov in *Science,* **158**, 946-947, 1967).

[2] Artsutanov, Y., "Into the Cosmos without Rockets", *Znanije-Sila* **7**, 25, 1969.

[3] Pearson, J., "The Orbital Tower: A Spacecraft Launcher Using the Earth's Rotational Energy", *Acta Astronautica* **2**, 785-799, 1975.

[4] Edwards, B. C., and Westling, E. A., "The Space Elevator: A Revolutionary Earth-to-Space Transportation System", ISBN 0972604502, published by the authors, January 2003.

[5] Mason, L. S., "A Solar Dynamic Power Option for Space Solar Power", Technical Memorandum NASA/TM—1999-209380 SAE 99–01–2601, 1999.

[6] Wyrsch, N. & 8 co-authors (2006) "Ultra-Light Amorphous Silicon Cell for Space Applications," Presented at 4th World Conference and Exhibition on Photovoltaic Solar Energy Conversion, March 2006, Waikoloa, Hawaii.

[7] S. Iijima, "Helical microtubules of graphitic carbon", *Nature* 56, 354 (1991).

[8] Ruoff et al., "Mechanical properties of carbon nanotubes: theoretical predictions and experimental measurements", *C. R. Physique* 4 [2003].

[9] T. Belytschko et al, "Atomistic Simulations of Nanotube Fracture", *PHYSICAL REVIEW B* [2002].

[10] B. Shelef, "Space Elevator Power System Analysis and Optimization", *The Spaceward Foundation*, 2008.

[11] B. Shelef, "Space Elevator Calculation Scrapbook", *The Spaceward Foundation*, 2008.

ADDITIONAL READING

The true father of the modern-day concept of the Space Elevator is Russian Engineer Yuri N. Artsutanov. In his own words, this is how he came to this idea:

> I was always interested in space and my friend Alik Yezrielev's father, as a Stalin Prize winner had access to foreign scientific and technical journals (that was in 1957, four years after Stalin's death) and we could also read them. On one occasion we read about a newly developed extremely strong polymer, so strong that if you used it to make a rope 400km in length it would not break under its own weight... The question arose what thickness would a rope of infinite length require. It turned out to be impossible if it was of a constant diameter. However, such a rope could be possible if it had a variable cross section, that is, was spindle shaped and if it was possible to use centrifugal force to counter the force of gravity. Step by step the idea of a lift into space was born. I kept talking to people about the idea but didn't submit my article to the Soviet newspaper "Komsomolskaya Pravda" until 1960 and a week later they published it[1]

This article was printed in Russian, of course, and was largely unknown to the western scientific community. Others, most notably American Engineer Jerome Pearson, unaware of Yuri's initial publication, independently reinvented the idea.

Yuri's paper remained obscure until the efforts of Mr. Roger Gilbertson. In the following pages, Roger chronicles his efforts to uncover and translate Yuri's work. Following that are reproductions of Yuri's original article and its translation into English.

All of us in the Space Elevator community are very grateful to Roger for his efforts in finding this article, having it translated into English and making both documents available world-wide.

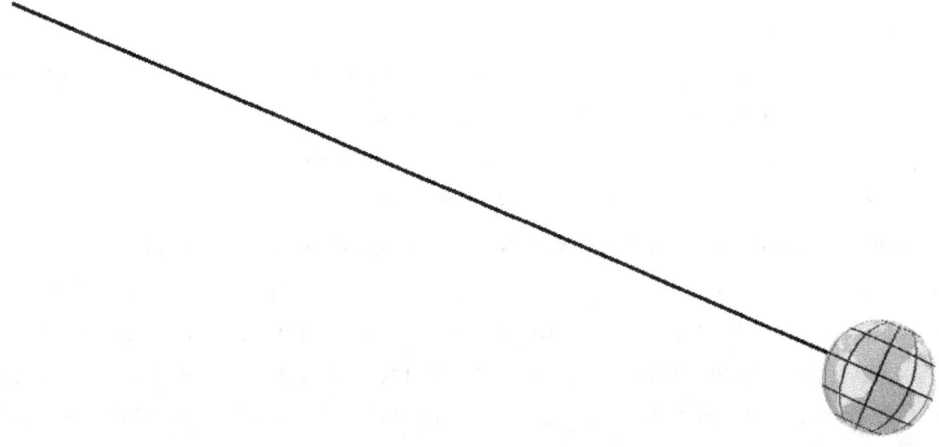

---

[1] From Conversations with Yuri Artustanov – Part III (http://www.spaceelevatorblog.com/?p=1408)

By Roger G. Gilbertson
rogermondo@yahoo.com

Like many, I first learned of the space elevator concept from Arthur C. Clarke's science fiction novel "Fountains of Paradise" first published in 1979. In 2003, after reading about Dr. Brad Edwards' NASA-funded study that concluded that the recently discovered carbon nanotubes could possibly allow the creation of a working space elevator, I became curious of about the origins of the idea.

I'd read mentions of Russian engineer Yuri Artsutanov and his 1960 article in the newspaper of the Soviet Union's communist youth organization *Komsomolskaya Pravda*, but could not locate the original story anywhere on the Internet. Eventually I learned that Stanford University's Hoover Institution possessed a complete collection of *KP* on microfilm. I found a librarian there, Ms. Molloy, eager to help. After a brief search, she sent me several high-contrast prints of the microfilmed Russian article.

I do not know Russian, so assembling the pages provided the first challenge. Eventually I realized that the pages formed a single page with three columns. I scanned the merged prints into my computer then, using a free demonstration version of a Cyrillic optical character recognition program called *FineReader 5 Pro* from Abbyy Software, I converted the article into a Macintosh document using a Cyrillic font. Proofreading the text file took a while, checking the software's output against the grainy microfilm prints, letter-by-letter – "double-K, upside-down L, backwards R…"

Finally, I emailed the Russian text file to my cousin-in-law, Joan, a Russian scholar at Catholic University in Washington, DC. She performed the bulk of the English translation, and I helped with choosing an occasional technical term. At last I could read the original article that had sparked the space elevator idea. Across the decades, Artsutanov's poetic phrases still inspired visions of a new and different way to leave the Earth and head for the stars. Now that I finally could read the original article, I wanted to share it with others.

Text files at hand, I laid out Russian and English versions that duplicated as closely as possible the original *Komsomolskaya Pravda* article. I included the artwork from the page, and then output PDF files of both. Finally, I sent the PDF files out to four or five Space Elevator web sites, and from there it has spread to quite a few places. I wanted to make the source material available in an easy-to-use form, and it has been enjoyable to watch the files find their way around the Internet.

About a year after releasing the PDF files, I received an email from Slava Rotkin, a researcher of fullerenes, nanotubes and related topics, originally from St. Petersburg, Russia. He sent a copy of my Russian PDF file that included notes pointing out a number of spelling corrections. Concerned that I'd made errors in the conversion or layout process, I returned to the original photocopy pages. But in fact, the typos were in the original 1960 document.

By being involved with the various Space Elevator games produced by the Spaceward Foundation, I got to know many of the players in the field. In 2005 I put together a proposal for a documentary film, and realized it would be exciting to film a reunion meeting of Clarke and Artsutanov (they'd first met in the 1970's). We made contact with Yuri, still living in St. Petersburg, and learned that he was well and able to travel. Unfortunately with Sir Arthur's passing in 2008, that opportunity was missed.

However with the August 2010 Space Elevator Conference in Seattle, I finally managed to meet Mr. Artsutanov in person. Through his translator, Eugene Schlusser, we carried on a long and enjoyable conversation about space elevators, the old Soviet Union, and the trials and tribulations of his epic idea for an entirely different way to reach the stars.

-- Roger G –

Yuri Artsutanov (left) and Roger Gilbertson meet for the first time at the 2010 Space Elevator Conference in Redmond, Washington. Photo: Natalie Gilbertson.

Roger G. Gilbertson lives and works in Los Angeles, California.

This is a completely revised and updated version of an article first published in
The Citizen Scientist in 2005.

(Link to PDF versions of original Komsomolskaya Pravda article and English translation)

http://digitaltimewarp.com/Writing/Entries/2005/11/18_Entry_1_files/Artsutanov_Pravda_SE.pdf

В космос — НА ЭЛЕКТРОВОЗЕ

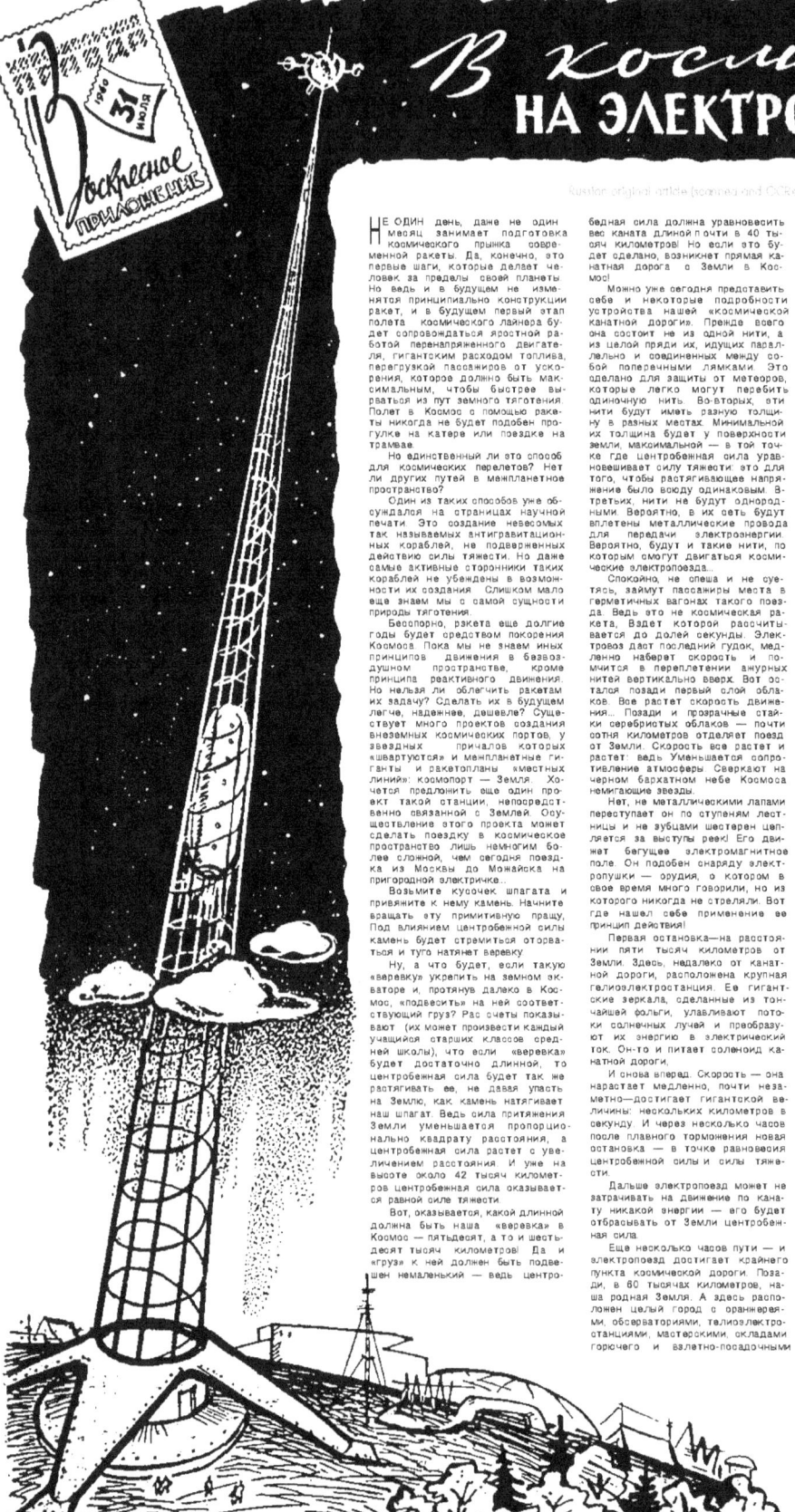

НЕ ОДИН день, даже не один месяц занимает подготовка космического прыжка современной ракеты. Да, конечно, это первые шаги, которые делает человек за пределы своей планеты. Но ведь и в будущем не изменятся принципиально конструкции ракет, и в будущем первый этап полета космического лайнера будет сопровождаться яростной работой перенапряженного двигателя, гигантским расходом топлива, перегрузкой пассажиров от ускорения, которое должно быть максимальным, чтобы быстрее вырваться из пут земного тяготения. Полет в Космос с помощью ракеты никогда не будет подобен прогулке на катере или поездке на трамвае.

Но единственный ли это способ для космических перелетов? Нет ли других путей в межпланетное пространство?

Один из таких способов уже обсуждался на страницах научной печати. Это создание невесомых так называемых антигравитационных кораблей, не подверженных действию силы тяжести. Но даже самые активные сторонники таких кораблей не убеждены в возможности их создания. Слишком мало еще знаем мы о самой сущности природы тяготения.

Бесспорно, ракета еще долгие годы будет средством покорения Космоса. Пока мы не знаем иных принципов движения в безвоздушном пространстве, кроме принципа реактивного движения. Но нельзя ли облегчить ракетам их задачу? Сделать их в будущем легче, надежнее, дешевле? Существует много проектов создания внеземных космических портов, у звездных причалов которых «швартуются» и межпланетные гиганты и ракетопланы «местных линий»: космопорт — Земля. Хочется предложить еще один проект такой станции, непосредственно связанной с Землей. Осуществление этого проекта может сделать поездку в космическое пространство лишь немногим более сложной, чем сегодня поездка из Москвы до Можайска на пригородной электричке..

Возьмите кусочек шпагата и привяжите к нему камень. Начните вращать эту примитивную пращу. Под влиянием центробежной силы камень будет стремиться оторваться и туго натянет веревку.

Ну, а что будет, если такую «веревку» укрепить на земном экваторе и, протянув далеко в Космос, «подвесить» на ней соответствующий груз? Рас счеты показывают (их может произвести каждый учащийся старших классов средней школы), что если «веревка» будет достаточно длинной, то центробежная сила будет так же растягивать ее, не давая упасть на Землю, как камень натягивает наш шпагат. Ведь сила притяжения Земли уменьшается пропорционально квадрату расстояния, а центробежная сила растет с увеличением расстояния. И уже на высоте около 42 тысяч километров центробежная сила оказывается равной силе тяжести.

Вот, оказывается, какой длинной должна быть наша «веревка» в Космос — пятьдесят, а то и шестьдесят тысяч километров! Да и «груз» к ней должен быть подвешен немаленький — ведь центробедная сила должна уравновесить вес каната длиной почти в 40 тысяч километров! Но если это будет сделано, возникнет прямая канатная дорога с Земли в Космос!

Можно уже сегодня представить себе и некоторые подробности устройства нашей «космической канатной дороги». Прежде всего она состоит не из одной нити, а из целой пряди их, идущих параллельно и соединенных между собой поперечными лямками Это сделано для защиты от метеоров, которые легко могут перебить одиночную нить. Во-вторых, эти нити будут иметь разную толщину в разных местах. Минимальной их толщина будет у поверхности земли, максимальной — в той точке, где центробежная сила уравновешивает силу тяжести эти для того, чтобы растягивающее напряжение было всюду одинаковым. В-третьих, нити не будут однородными. Вероятно, в их сеть будут вплетены металлические провода для передачи электроэнергии. Вероятно, будут и такие нити, по которым смогут двигаться космические электропоезда.

Спокойно, не спеша и не суетясь, займут пассажиры места в герметичных вагонах такого поезда. Ведь это не космическая ракета. Взлет которой рассчитывается до долей секунды. Электровоз дает последний гудок, медленно набирает скорость и помчится в переплетении ажурных нитей вертикально вверх. Вот остался позади первый слой облаков. Все растет скорость движения... Позади и прозрачные стайки серебристых облаков — почти сотня километров отделяет поезд от Земли. Скорость все растет и растет: ведь уменьшается сопротивление атмосферы. Сверкают на черном бархатном небе Космоса немигающие звезды.

Нет, не металлическими лапами переступает он по ступеням лестницы и не зубцами шестерен цепляется за выступы реек! Его движет бегущее электромагнитное поле. Он подобен снаряду электропушки — орудия, о котором в свое время много говорили, но из которого никогда не стреляли. Вот где нашел себе применение ее принцип действия!

Первая остановка—на расстоянии пяти тысяч километров от Земли. Здесь, недалеко от канатной дороги, расположена крупная гелиоэлектростанция. Ее гигантские зеркала, сделанные из тончайшей фольги, улавливают потоки солнечных лучей и преобразуют их энергию в электрический ток. Он-то и питает всю канатной дороги.

И снова вперед. Скорость — она нарастает медленно, почти незаметно—достигает гигантской величины: несколько километров в секунду. И через несколько часов после плавного торможения новая остановка — в точке равновесия центробежной силы и силы тяжести.

Дальше электропоезд может не затрачивать на движение по канату никакой энергии — его будет отбрасывать от Земли центробежная сила.

Еще несколько часов пути — и электропоезд достигает крайнего пункта космической дороги. Позади, в 60 тысячах километров, наша родная Земля. А здесь расположен целый город с оранжереями, обсерваториями, гелиоэлектростанциями, мастерскими, складами горючего и взлетно-посадочными устройствами для межпланетных ракет. Нет, живущие здесь люди не оторваны от Земли. Они накрепко связаны с ней в самом прямом смысле нитью космической дороги..

Отправляющиеся со здешнего космодрома ракеты совсем не похожи на те, что, грохоча взрывами, взлетают с Земли. Ведь здесь они уже имеют космическую скорость, вместе с космодромом вращаясь вокруг Земли. Здесь нет тяготения, которое заставляет делать земные ракеты массивными и прочными. Здесь не нужны сверхмощные двигатели. Космические ракеты плавно покидают причальные сооружения и подходят к ним, похожие в своей неспешной неповоротливости на океанские суда...

Такой нам представляется сегодня эта космическая дорога.

Самое сложное в ее строительстве, вероятно, самое начало. Для этого надо будет забросить в равновесную зону искусственный спутник, на котором будет находиться в собранном виде первая нить — в минимальном сечении тоньше человеческого волоса. И то ее вес окажется около тысячи тонн. И с этого спутника надо будет спускать сразу два конца этой дороги: один — на Землю, другой — о космическое пространство.

Когда первая нить будет закреплена на Земле, используя ее как опору, можно пустить по ней автоматического «паучка», который потянет вторую параллельную нить, затем третью, четвертую и т д.

Космические канатные дороги можно создать и на некоторых других планетах и спутниках Очень медленно вращается наша Луна, невелика поэтому центробежная сила, вызываемая ее вращением. Но ведь уже на расстоянии 57 тысяч километров от поверхности нашего вечного спутника сила притяжения Земли начинает превосходить силу лунного притяжения. Значит, укрепив канат в центре видимого лунного диска, протянув его в направлении к Земле на расстояние больше 57 тысяч километров и уравновесив соответствующим грузом, мы получим отличную лунную канатную дорогу. А построив эти две дороги, можно будет осуществлять переезд Земля — Луна почти без расходования топлива.

От Меркурия, который все время обращен одной стороной к Солнцу, легко «опустить» канат в сторону нашего центрального светила Устроить канатную дорогу на Марсе значительно легче, чем на Земле—ведь его притяжение значительно меньше земного. Трудно сказать что-либо о загадочной Венере, период вращения которой определен весьма неточно. Вероятно, можно будет соорудить такие дороги и на многих спутниках крупных планет...

Но, конечно, все это — дело далекого будущего. Чтобы осуществить его, надо преодолеть множество препятствий. Нет еще материала, прочность которого могла бы выдержать гигантский вес канатной дороги с Земли в Космос. Самые прочные пластмассы и стали в несколько раз слабее, чем требуется. Вот лунную канатную дорогу уже можно было бы строить—ведь для нее нужен значительно менее прочный материал.

Надо тщательно изучить и многие другие вопросы: влияние вибраций, поведение различных веществ под влиянием космического излучения и т. д.

Но наука и техника стремительно движутся вперед, и, может быть, еще в пределах нашего века начнется сооружение канатной дороги на небо.

Ю. АРЦУТАНОВ,
аспирант Ленинградского технологического института.

Рис. А. ГУРЬЕВА.

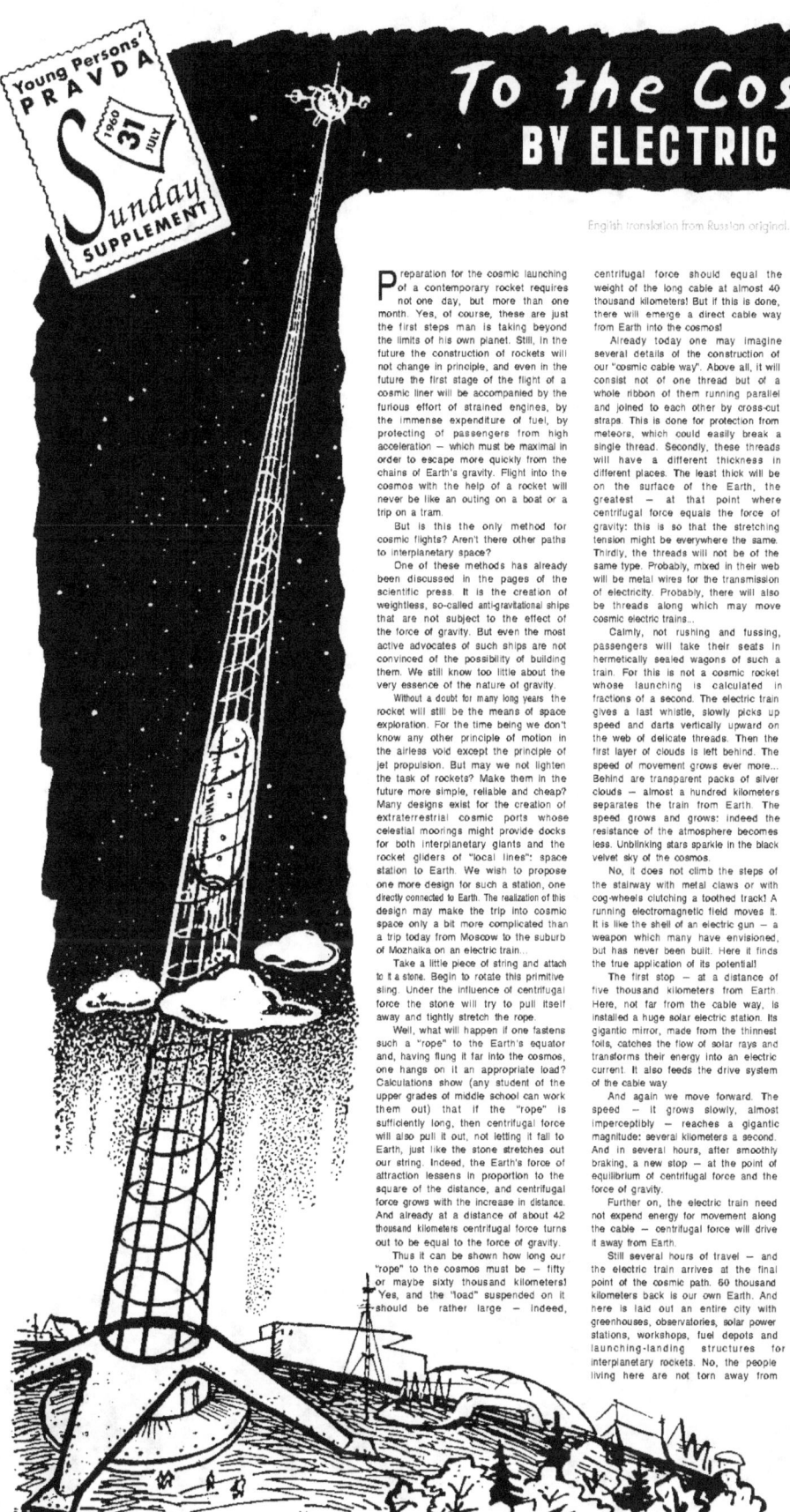

Young Persons' PRAVDA

Sunday SUPPLEMENT

1960 31 JULY

To the Cosmos
BY ELECTRIC TRAIN

English translation from Russian original.

Preparation for the cosmic launching of a contemporary rocket requires not one day, but more than one month. Yes, of course, these are just the first steps man is taking beyond the limits of his own planet. Still, in the future the construction of rockets will not change in principle, and even in the future the first stage of the flight of a cosmic liner will be accompanied by the furious effort of strained engines, by the immense expenditure of fuel, by protecting of passengers from high acceleration — which must be maximal in order to escape more quickly from the chains of Earth's gravity. Flight into the cosmos with the help of a rocket will never be like an outing on a boat or a trip on a tram.

But is this the only method for cosmic flights? Aren't there other paths to interplanetary space?

One of these methods has already been discussed in the pages of the scientific press. It is the creation of weightless, so-called anti-gravitational ships that are not subject to the effect of the force of gravity. But even the most active advocates of such ships are not convinced of the possibility of building them. We still know too little about the very essence of the nature of gravity.

Without a doubt for many long years the rocket will still be the means of space exploration. For the time being we don't know any other principle of motion in the airless void except the principle of jet propulsion. But may we not lighten the task of rockets? Make them in the future more simple, reliable and cheap? Many designs exist for the creation of extraterrestrial cosmic ports whose celestial moorings might provide docks for both interplanetary giants and the rocket gliders of "local lines": space station to Earth. We wish to propose one more design for such a station, one directly connected to Earth. The realization of this design may make the trip into cosmic space only a bit more complicated than a trip today from Moscow to the suburb of Mozhaika on an electric train...

Take a little piece of string and attach to it a stone. Begin to rotate this primitive sling. Under the influence of centrifugal force the stone will try to pull itself away and tightly stretch the rope.

Well, what will happen if one fastens such a "rope" to the Earth's equator and, having flung it far into the cosmos, one hangs on it an appropriate load? Calculations show (any student of the upper grades of middle school can work them out) that if the "rope" is sufficiently long, then centrifugal force will also pull it out, not letting it fall to Earth, just like the stone stretches out our string. Indeed, the Earth's force of attraction lessens in proportion to the square of the distance, and centrifugal force grows with the increase in distance. And already at a distance of about 42 thousand kilometers centrifugal force turns out to be equal to the force of gravity.

Thus it can be shown how long our "rope" to the cosmos must be — fifty or maybe sixty thousand kilometers! Yes, and the "load" suspended on it should be rather large — indeed,

centrifugal force should equal the weight of the long cable at almost 40 thousand kilometers! But if this is done, there will emerge a direct cable way from Earth into the cosmos!

Already today one may imagine several details of the construction of our "cosmic cable way". Above all, it will consist not of one thread but of a whole ribbon of them running parallel and joined to each other by cross-cut straps. This is done for protection from meteors, which could easily break a single thread. Secondly, these threads will have a different thickness in different places. The least thick will be on the surface of the Earth, the greatest — at that point where centrifugal force equals the force of gravity: this is so that the stretching tension might be everywhere the same. Thirdly, the threads will not be of the same type. Probably, mixed in their web will be metal wires for the transmission of electricity. Probably, there will also be threads along which may move cosmic electric trains...

Calmly, not rushing and fussing, passengers will take their seats in hermetically sealed wagons of such a train. For this is not a cosmic rocket whose launching is calculated in fractions of a second. The electric train gives a last whistle, slowly picks up speed and darts vertically upward on the web of delicate threads. Then the first layer of clouds is left behind. The speed of movement grows ever more... Behind are transparent packs of silver clouds — almost a hundred kilometers separates the train from Earth. The speed grows and grows: indeed the resistance of the atmosphere becomes less. Unblinking stars sparkle in the black velvet sky of the cosmos.

No, it does not climb the steps of the stairway with metal claws or with cog-wheels clutching a toothed track! A running electromagnetic field moves it. It is like the shell of an electric gun — a weapon which many have envisioned, but has never been built. Here it finds the true application of its potential!

The first stop — at a distance of five thousand kilometers from Earth. Here, not far from the cable way, is installed a huge solar station. Its gigantic mirror, made from the thinnest foils, catches the flow of solar rays and transforms their energy into an electric current. It also feeds the drive system of the cable way.

And again we move forward. The speed — it grows slowly, almost imperceptibly — reaches a gigantic magnitude: several kilometers a second. And in several hours, after smoothly braking, a new stop — at the point of equilibrium of centrifugal force and the force of gravity.

Further on, the electric train need not expend energy for movement along the cable — centrifugal force will drive it away from Earth.

Still several hours of travel — and the electric train arrives at the final point of the cosmic path. 60 thousand kilometers back is our own Earth. And here is laid out an entire city with greenhouses, observatories, solar power stations, workshops, fuel depots and launching-landing structures for interplanetary rockets. No, the people living here are not torn away from

Earth. They are firmly tied to it in a most direct sense by the thread of the cosmic path...

The rockets setting out from the space station here are in no way similar to those that, roaring with explosions, lift off from Earth. Indeed, here they already have cosmic speed, together with their space station rotating around Earth. Here there is not the gravity that compels one to make Earth rockets massive and solid. Here super powerful engines are not needed. Cosmic rockets smoothly leave their mooring structures and return to them, similar in their unhurried pace to ocean liners...

So this cosmic tram way seems to us today.

The most complicated aspect of its construction is, probably, the very beginning. For this it will be necessary to place into geosynchronous orbit an artificial satellite on which the first thread will be located in assembled form — in its smallest section thinner than a human hair. And yet its weight will turn out to be about a thousand tons. And from this satellite it will be necessary to let down out simultaneously the two ends of our pathway: one towards Earth, the other towards cosmic space.

When the first thread is fastened to Earth, using it as a pier, one may send along it an automated "spider" which will haul a second parallel thread, and then a third, fourth, and so on.

One may also create cosmic cable ways on several other planets and satellites. Our Moon rotates very slowly, thus the centrifugal force elicited by its rotation is not great. But at a distance of 57 thousand kilometers from the surface of our eternal satellite, the Earth's force of attraction begins to exceed the force of lunar attraction. This means that, once the cable is fastened in the center of the visible lunar disk, pulled in the direction of Earth to a distance of more than 57 thousand kilometers and balanced with an appropriate load, we will get an excellent lunar cable way. And having built these two paths, it will be possible to achieve an Earth-Moon route traversable almost without the expenditure of fuel.

From Mercury, which is always turned with one side toward the Sun, it is easy to "let down" a cable on the side of our central luminary. To build a cable way on Mars is significantly easier than on Earth — for its gravity is significantly less than Earth's. It is difficult to say anything about the mysterious Venus, whose period of rotation is extremely ill-defined. Probably it will be possible to erect such paths on many of the satellites of the large planets...

Well, of course, this is all a matter for the distant future. To realize it, one must overcome a multitude of obstacles. There is still no material whose durability could bear the gigantic weight of a cable way from Earth to the cosmos. The most durable plastics and steels are several times weaker than required. Though it might already be possible to build a lunar cable way since significantly less durable material is needed for that.

It is also necessary to study carefully many other questions: the influence of vibrations, the behavior of various substances under the influence of cosmic radiation, and so forth.

But science and technology are swiftly moving ahead, and, perhaps, already toward the end of our century the construction of a cable way to the heavens will begin.

YU. ARTSUTANOV,
aspirant, Leningrad
Technological Institute.

Illus. A. Guryeva.

SEVEN DEADLY ASSUMPTIONS ABOUT SPACE ELEVATORS

Gaylen Hinton
gaylenhinton@yahoo.com

Abstract: There are many space elevator (SE) concepts that have been widely accepted as "gospel truth" that are not necessarily so. This paper addresses seven of these assumptions and mind-sets that may have hindered the progress and development of SEs. These seven assumptions are as follows:

1. A space elevator must be built from carbon nanotubes.
2. There will be problems with oscillations and vibrations on a space elevator.
3. A space elevator needs to be tapered.
4. A space elevator will have climbers that go up and down the cable.
5. In order to avoid storms and lightning, an SE base station needs to be on a ship in a largely storm-free location.
6. A space elevator needs to be on a mobile base in order to avoid space junk.
7. A space elevator will be very expensive to construct.

Each of these assumptions will be discussed and refuted.

1.   Introduction

As any new science or technology develops, there are initial assumptions and beliefs that can become entrenched in the mind-sets of the people involved. Over time, these assumptions can become canonized and unchallenged as simply "the way it is". The same has been true for the development of the SE. This paper will address seven of these assumptions, and attempt to clarify some of the issues behind them.

2.   Assumption 1: A space elevator must be built from carbon nanotubes

Although carbon nanotubes have an impressive theoretical maximum strength, their real-world strength will probably never be good enough to make a practical SE. When the unavoidable nanotube defects are factored into the equation, there are many other substances whose real-world strength could far exceed that of carbon nanotubes. Some of these substances will be discussed.

Carbon nanotubes (CN) are essentially sheets of graphene that have been rolled up into a tube. Their strength is the strength of graphene. Actual graphene sheets have been tested to have a tensile strength of 130 GPa [1]. Some have calculated that graphene could have a theoretical strength of up to 300 GPa [2].

Therefore CN could have a theoretical strength that is well over 100 GPa, which is more than enough to build a good space elevator. Yet that anticipated strength has been unachievable in real-world CN tests. Defects and imperfections in the CN are the culprit.

One of the problems is that a chain is only as strong as its weakest link. Any defect or imperfection in a long CN will cause it to break at that point. That means that hundreds of millions (if not billions) of atoms need to be correctly assembled in each CN segment. Given the many and varied ways that carbon atoms can and will bond, that is a very tall order.

Even if defect-free CN were able to be produced initially, they would not stay that way for very long. There is a spontaneous defect that occurs in CN called the *Stone Wales defect*. That defect weakens the CN and limits the ultimate tensile strength to a value that would only be marginally good enough for making an SE.

Other materials are not limited by the Stone Wales defects, or even with defects in the initial structure like CN are. Therefore they could have a real-world tensile strength much greater than that of CN.

CN have a three dimensional structure, whereas a simple molecular chain essentially has a one-dimensional structure. There are a lot more defects that can occur in building a three-dimensional structure than with a one-dimensional structure.

In addition, defects in three dimensional structures produce stress concentrations, which can tear the material apart. Stress concentrations are what limit the ultimate strength of all three dimensional materials. A 1% defect does not make the material lose 1% of its strength. A 1% defect causes a stress concentration that makes the material lose 99% or more of its strength.

However, in a one dimensional structure there cannot be significant stress concentrations. Perhaps we should be looking for one-dimensional materials to build a space elevator.

One of the simplest possible one-dimensional materials that could have a greater real-world strength than CN is the common plastic polyethylene (PE) – the material of garbage bags.

While grocery bags and milk jugs hardly seem to possess any kind of space-elevator strength, it is important to realize that when these materials rupture, it is *not* the atomic bonds that are breaking. The polymer chains are merely slipping past one another. As the molecular chains get longer and longer, they slip less and less, resulting in a greater strength.

The commercial fibers *Spectra* and *Dyneema* are PE fibers. The length of the PE molecules in those fibers is long enough to give those fibers the strength to make bullet-proof vests. Those PE fibers have one of the highest specific strengths of any commercial fibers. Yet they still exhibit *creep* when they are heavily loaded, meaning that the molecular chains are still slipping, rather than breaking. They have not yet reached the maximum strength of the PE molecule.

In Fig. 1A below the molecular structure of graphene (or CN) is shown. In Fig 1B the structure of the carbon atoms in a few PE molecules are shown.

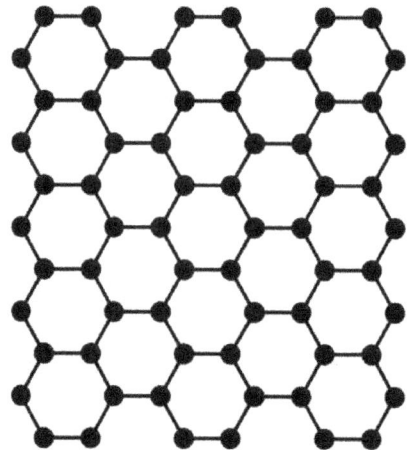

Figure 1A Graphene

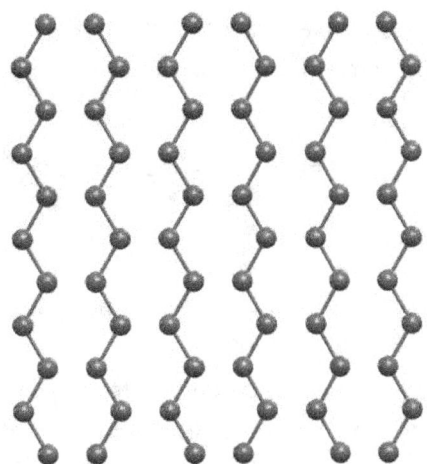

Figure 1B Polyethylene

As can be seen by looking at these two figures, the carbon atom structure in the vertical direction is *exactly the same* for both. The only difference is that the carbon atoms are bound together in the horizontal direction in graphene. Those sideways carbon bonds (which provide no vertical strength) are replaced with hydrogen atoms in the polyethylene.

The load bearing structures of polyethylene and graphene are *identical*. Therefore, it would also be the same as CN.

Because of the addition weight of the hydrogen atoms, and the fact that the carbon atoms are not quite as close, the specific strength of the PE would be about 70% of the maximum theoretical specific strength of the graphene. That means that PE fibers could still end up with a specific strength of 100 MN-m/kg, which would be much higher than that of real-world CN fibers. PE could make a great SE.

It should be noted that with PE fibers, there can be no Stone Wales defects. There also are virtually no imperfections in the molecular chain – either you have the chain or not. If the chains are long enough, they will be essentially defect-free.

PE chains are made through a variety of catalysts that are very selective in creating defect free PE chains. In fact, effort has to be made to keep the chains *short enough* in commercial PE production. Hydrogen is added, and the operating temperature is adjusted so as to force the chains to be shorter.

It is very easy to make PE chains that are so long that they have *no commercial value*. That is because as the chains get longer, the melting temperature rises until the PE will no longer melt, but simply decomposes as the temperature gets too hot.

Spectra and *Dyneema* fibers are made from the longest PE molecules that can be solvent-extracted and made into fibers. They are not the longest PE molecular chains that can be made, but simply the longest molecular chains that can currently be *processed*.

With the selective catalysts doing the work, it is easy to make very long, defect free PE molecules. The problem is that those long PE molecules tend to curl up like a ball of yarn, and cannot be melted or extracted using current technologies.

Therefore, the real technological need is not in how to make long, defect-free PE molecules, but how to get them straightened out and lined up in an SE cable.

Another likely candidate for an SE cable is *polycumulene*. Polycumulene is simply a linear chain of carbon atoms with consecutive double bonds as shown below:

$$\ldots C=C=C=C=C=C=C=C=C=C=C \ldots$$

Although short double-bonded chains of several carbon atoms can be unstable, as the chain length increases beyond a certain number, the stability increases. It has been calculated that a polycumulene chain with more than about 3000 atoms would be stable at room temperature [3].

A polycumulene chain long enough to be spun into a macroscopic fiber would be extremely stable. Polycumulene has been found on meteorites and has been detected in space.

Because of the double bond, the tight atomic spacing, and the linear construction, a polycumulene chain would have an even higher specific strength than the theoretical maximum of graphene or CN. A linear carbon sample (which was likely polycumulene) was tested in 2005 to have a Young's Modulus forty times higher than that of diamond, the hardest commercial material [4].

Some other compounds that could make macroscopic fibers stronger than CN are shown below, but this list is certainly not complete:

$$
\begin{array}{ccccccc}
H & & H & & H & & H \\
| & & | & & | & & | \\
\ldots N & - N - N & - N - N & - N - N & - N \ldots \\
| & & | & & | & & | \\
H & & H & & H & & H
\end{array}
$$

Polyhydrazine

$$\ldots N=N-N=N-N=N-N=N \ldots$$

Polyazide

$$
\begin{array}{c}
H\ HHH\ HHH\ HHH\ HH \\
|\ \vee\ |\ \vee\ |\ \vee\ |\ \vee \\
\ldots B-B-B-B-B-B-B-B \ldots \\
\wedge\ |\ \wedge\ |\ \wedge\ |\ \wedge\ | \\
HHH\ HH\ HHHHHH\ H
\end{array}
$$

Polyborane

3. Assumption 2: There will be problems with oscillations and vibrations on an SE

In space there is no air resistance to dampen oscillations. Therefore, there have been fears that oscillations could be amplified by a resonance until the SE was destroyed or damaged.

In a space elevator there are three modes of oscillation, and we will analyze the impacts of each of these modes on the SE:

3.1 Longitudinal mode:

This is the stretching mode, like a ball bouncing on the end of a rubber band. This mode is not going to be an issue because any vibrations will be quickly damped out, even though there is no air resistance.

A bouncing weight on a rubber band is always quickly damped out, even in a vacuum. The vibration is damped by the heating of the stretched band, not the air resistance. Therefore any longitudinal vibrations on the SE will be damped out by the heat dissipated in the cable itself.

3.2 Pendulum mode:

This is where the whole SE swings like a pendulum. In this case, there is no air resistance and the amount of energy dissipated in the cable itself as a result of the pendulum motion is negligible. However, in the real world this oscillation mode will not be a serious problem either.

First of all, there is no such thing as a pure pendulum motion in an SE. Any pendulum motion will also excite a stretching oscillation due to the changing force on the cable. That longitudinal oscillation will dissipate energy, creating a pendulum damping effect – even in a vacuum.

However, even if we completely ignore any damping effects caused by the cable stretching, there still would not be a problem with pendulum oscillations on an SE. The calculations and discussions that follow totally ignore any damping effects, yet they still show that the pendulum oscillations would not be a problem.

Even if there were zero damping on a space pendulum, there would still have to be a driving force to cause the pendulum's swing to increase. If that driving force is exactly in resonance with the oscillation then the maximum excursions will increase. However, if the driving force is *not* at the pendulum's resonance, then the maximum excursions can only increase to a limiting value, even with no damping.

The equation for the maximum force (which is proportional to the maximum amplitude) in a harmonic oscillator is:

$$F_{max} = \frac{F_0}{\sqrt{\left(1 - \left(\frac{\omega}{\omega_0}\right)^2\right)^2 + \left(\frac{2\zeta\omega}{\omega_0}\right)^2}} \tag{1}$$

Where F_0 is the driving force,

ω is the driving frequency,

ζ is the damping ratio, and

ω_0 is the resonant freqency

In the vacuum of space we are assuming that the damping ratio ζ is zero, so the above equation simplifies to:

$$F_{max} = \frac{F_0}{\left| 1 - \left(\frac{\omega}{\omega_0}\right)^2 \right|} \tag{2}$$

A graph of the maximum force of equation (2) with zero damping is shown in Fig. 2 below in the vicinity of $\omega = \omega_0$.

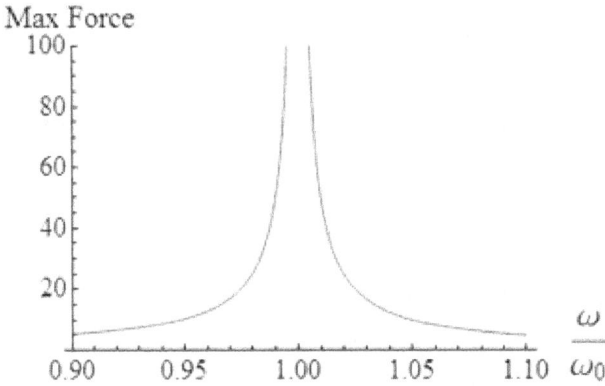

Max Force

Figure 2. Resonant forces

Obviously, very close to the resonance frequency, the driving force can be greatly amplified. However, if the driving frequency is even slightly removed from the resonance frequency, the amplification is severely limited.

In effect, an off-resonant driving frequency has an *effective damping ratio*, ζ_{eff}, at that frequency:

$$\zeta_{eff} = \frac{1}{2F_{max}}$$

ζ_{eff} would give the same results as an oscillator with a real damping ratio, ζ, that was being driven at its resonant frequency. The physical explanation of ζ_{eff} is that the driving frequency will be out of phase with the resonance for part of the time, and would be fighting against the oscillations, keeping them at a low level.

Let us look at a specific example. One of the most obvious driving forces that could excite a pendulum mode of oscillation in an SE is the gravity of the moon. It would affect the SE every 24.81 hours as the rotation of the earth caused it to pass the moon.

Let us assume that we built an SE that had a counterweight of 500,000 kg at 110,000 km. The force on an SE counterweight is:

$$F = m\omega^2 r - \frac{mGM_E}{r^2}$$

Solving for F gives a force of 276,000 N, which gives an effective acceleration of 0.552 m/s^2.

Now, the period of a pendulum is:

$$T = 2\pi \sqrt{\frac{l}{a}}$$

With a length of 110,000 km and an effective acceleration of 0.552, we get a period of 24.638 hrs.

That resonance appears to be close enough to the period of the moon's interaction to cause a serious problem – it is less than 1% off. However, if we calculate the maximum force from equation (2) above, we see that even with no damping at all in space, the effect of the moon's gravity could never be amplified by more than a factor of 72.

At 110,000 km from the earth, the counterweight would be about 270,000 km from the moon. The maximum force from the moon's gravity on the counterweight would be:

$$F = \frac{mGM_M}{r^2}$$

Plugging in the numbers from the counterweight gives a force of only 34 N. Multiplying that by the amplification factor of 72 gives the maximum force that could ever be on the SE pendulum: 2400 N. That is less than one hundredth of the initial tension that the counterweight is providing, and so would be insignificant.

Therefore, even though the frequency of the passing moon was within 1% of the resonant frequency of the SE, it could never move the SE more than a half of a degree in a pendulum motion.

Even if someone were foolish enough to construct an SE whose pendulum frequency was exactly in resonance with the moon, there would still not be a serious problem. That is because the actual pendulum frequency changes with amplitude.

The restoring force on a pendulum as it swings is proportional to $\sin(\theta)$. However, for small angles of swing, we usually assume $\sin(\theta) = \theta$. That assumption makes it easy to deal with mathematically, and gives us a simple harmonic oscillator. However, as the pendulum's swing increases, $\sin(\theta) \neq \theta$, and the harmonic oscillator assumption no longer applies.

As the angle of swing of a pendulum increases, the period of oscillation increases according to the following formula:

$$T = T_0 \left(1 + \frac{\theta^2}{16} + \frac{11(\theta^4)}{3072} + higher\ order\ terms \right) \tag{3}$$

Therefore, as the pendulum's amplitude increases, its resonant frequency changes. What this means is that, even if there *were* a driving force that was exactly in resonance with the pendulum at lower amplitudes, it would be off- resonance at higher amplitudes, and would be incapable of transferring more energy to the pendulum.

As the pendulum swings to its maximum extent, we have the following relations as shown in Fig. 3 below:

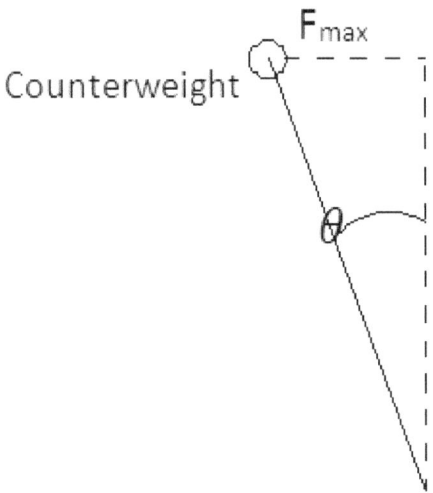

Figure 3. Swinging space elevator

$$\frac{F_{max}}{Initial\ tension} = Tangent(\theta)$$

So, for our SE we would have, using equation (2):

$$\frac{F_{max}}{276{,}000} = tan(\theta) = \frac{F_0}{276{,}000\left(1 - \left(\frac{\omega}{\omega_0}\right)^2\right)}$$

But $F_0 = 34$ N, and ω/ω_0 changes with amplitude angle according to equation (3). So we would end up with the relation:

$$tan(\theta) = \frac{34}{276{,}000\left(1 - \left(1 + \frac{\theta^2}{16}\right)^2\right)}$$

Solving for θ, we get, θ = 5.7°. That is the maximum angular swing that the SE could ever make due to the gravity of the moon.

At that amplitude, the resonant frequency would have changed enough so that

$$F_{max} = 813 \times F_0 = 27600 \text{ N}$$

Those 27600 N of force would alternate between a centrifugal force when the SE was at the center of its swing, and a tangential force at the limits of the swing.

As *F = ma,* that implies that the maximum acceleration of this worst case SE pendulum would be:

$$a = \frac{F}{m} = \frac{27600 \ N}{500000 \ kg} = .055 \frac{m}{s^2} = \frac{1}{177} g$$

Such a small amount of *g* force would not even be noticeable to a person on the SE.

However, because the counterweight only feels a *g* force of about 1/18*g* to begin with, 1/177*g* is a 10% change. A 10% periodic change in initial tension would make a logistical problem of coordinating the lifting of SE cars of various loads.

In addition, the movement of the SE would have to be taken into account for any satellite releases.

Therefore, a large pendulum vibration on an SE would be an annoyance, but it would not be damaging.

It should be noted that it would take a long time for a moon resonance to start swinging the SE at 5.7°. A driven resonance oscillation increases in amplitude over time until it reaches a steady state condition, approximately according to the envelope:

$$Amplitude \propto (1 - e^{-\zeta \omega t})$$

In our case we would have:

$$\zeta_{eff} = \frac{1}{2F_{max}} \quad and \quad \omega t = 2\pi N$$

$$with \ N = number \ of \ cycles$$

So the amplitude of the oscillations would increase over time:

$$Max \ swing \ of \ SE = \ 5.7° \left(1 - e^{-\frac{2\pi N}{1626}}\right) \qquad\qquad (4)$$

According to equation (4) it would take 600 cycles to get out to about 90% of the maximum swing. Yet each oscillation cycle takes 24.81 hours to happen. That is 620 days for 600 cycles. It would literally take years for the SE to be slowly accelerated out to its maximum swing of 5.7° by the moon's gravity. And that is the very worst possible case -ignoring the reality of the damping caused by the stretching.

All of this means that a moon resonance could never cause a serious problem with an SE, and no other repetitive force could ever last long enough to have any significant effect on the pendulum mode. If the damping effects of the induced tether stretching are taken into account, then there is even less concern.

3.3   Vibrating string mode:

This is where there would be a variety of oscillations in the SE like vibrations on a guitar string. In space there is no damping in this mode either, and a driving force could amplify oscillations. Moving elevator cars, storms, and satellite releases could all excite this mode.

However, just like with the pendulum mode, any driving force that is off-resonance has a limited amplification potential. In addition, a vibrating string mode induces stretching in the SE tether - just like with the pendulum mode - which dissipates energy in that mode too. Also, any cyclic perturbation has to last for a very long time to have any significant effect.

The restoring force in the vibrating string mode is also proportional to sin(θ), so it is also *not* a simple harmonic oscillator. In fact, there are multiple approximations involved in treating a vibrating sting as a harmonic oscillator. Like the pendulum, the frequency also changes with amplitude, but its frequency changes even faster.

Therefore, no vibrating string mode of oscillation could ever be excited to the point of causing damage. Unless there was a significant repetitive driving force that lasted a very long time, a vibrating string mode of oscillation could never damage the SE.

Yet, there is no possible repetitive driving force that could ever impart enough force over a long enough time frame to cause a problem.

What all this means to the SE is that there will be many oscillations in the longitudinal, pendulum, and vibrating string modes. The SE will be flopping around in a variety of ways. But who cares?

None of these vibrations would be potentially damaging. If the SE was designed with a 15% safety factor to deal with any possible oscillation issues, it would always be safe. In addition, by the timing and speed of elevator cars and satellite releases, existing vibrations of the pendulum and vibrating string modes could be attenuated if needed.

Because of the great length of the SE, any vibrational frequencies will have a very long period, typically measured in hours. Therefore, even though the SE may be vibrating in several different modes simultaneously, any people on the SE would probably never even feel the vibrations. As the accelerations would be very small, and the movements would be slow and gentle, there would never be a problem

Even if a very large pulse was sent up the SE, for example from a near-fatal space junk collision, it would spread out as it propagated along the cable. *Any* pulse will always spread out as it propagates in a medium, particularly when there are multiple possible transmission modes. Due to the huge distances involved, a powerful pulse would only be perceived as a gentle rocking motion by the time it got very high on the SE.

Even if such a pulse had not spread out by the time it interacted with SE equipment, it would still not cause a problem. That is because the cable itself, as well as any mechanisms, connections, and hardware, would all be designed to handle the initial tension of the SE, which is a greater load than any pulse could ever impart. (Otherwise the pulse would break the cable.) If the pulse was not strong enough to break the tether, it would not be strong enough to cause a problem.

Although this discussion about SE vibrations is certainly very basic and simplistic, it shows that concerns about vibrational issues may not be as serious as some had assumed.

4. Assumption 3: A space elevator needs to be tapered.

Theoretically, a tapered SE looks good, as the stresses on all parts would be equal. However, in the real world, a tapered construction would be unnecessary, expensive, and very impractical to make.

In 1975 Jerome Pearson did a great job of laying out some of the mathematical foundations for the SE [5]. In that article he discussed the need to have a tapered SE to keep the stress constant throughout the total structure.

However, the stress is not constant in any other structure or product ever created by man. Why should the SE be any different? Uniform stress is a theoretical ideal, but a real-world impracticality.

The only realistic way for a taper to be created on an SE would require that all the rolls of cable material be sent up to or near geostationary orbit (GEO). Once there, the rolls of cable could be unrolled down the SE structure and secured with no relative movement. It would always have to be an unrolling process, as it would never be practical to have new cables being dragged past the existing tether.

Anyone who has ever seen two fishing lines get tangled together knows that one cable cannot be pulled past another cable. It is a fool's errand to ever try to have relative motion between two cables. All construction of an SE needs be an unrolling process.

However, such a construction method would not be practically using a seed cable, with the material being sent up that cable. Very small rolls of cable would have to be sent up the cable on climbers. Once they reached GEO, they would then be unrolled back down (or up) the SE. It would be an impractically slow method of construction, being a highly non-continuous process.

In addition, while the cable was being unrolled onto a tapered SE, it would have to be unrolled *under tension*. As the underlying tether would be under tension, any additional strands would have to have the same tension or they would provide no strength. In fact, if the additional strands were not at the same tension, they would add a load to the original tether instead of strength.

As there will be over 1000 km of stretch in an SE tether between earth and GEO, it is obvious that any additional strands have to be stretched a lot in order to have the same tension and provide added strength.

Unrolling a cable under high tension would be a very slow, energy expensive process using a climber. Using all of the energy that could be beamed to a climber, the tension requirements would force it to move much much slower than it could travel unloaded. That would vastly slow down the construction process of a tapered SE.

Waiting for a climber to go up, unroll its cable under tension, and then come back down would take a tremendous amount of time. Even if the climbers were used for counterweights or discarded after unrolling their cable, the climbers would still have to wait on each other. That is because each climber would have to get far out of the gravity well of the earth before another could start.

On the other hand, if cable could be fed continuously from earth to GEO, only 13.8% of the earthbound weight of that length of cable would be a load on the tether. In addition, a continuously fed cable could be fed out at least five times faster than a climber could climb. And, there would be no waiting for cable rolls to be unwound under tension. The net result is that an SE constructed by continuously fed cables could be finished twenty times faster than one made by climbers.

Unfortunately there is no continuous process of feeding cables from earth that could create a tapered SE.

If any one of the materials discussed above in *Assumption 1* becomes available, then the SE could support its own weight from GEO and no taper would be needed. The SE could be fully constructed with the seed-cable counterweight force pulling up the entire SE, as was discussed in the 2009 Space Elevator Conference [6]. There would be no need for climbers, movement, overlap, or bonding of cables, and the entire construction would be a continuous process of lifting new cables.

If the ultimate strength of the available material does not allow it to support its own weight from GEO, then a step SE could be made with two or more moving belts. This design and construction could be done with small seed cables to earth, and was also discussed in the 2009 Space Elevator Conference [7]. There would be no taper, and no pulling of cables past cables. This construction method would also be a continuous process.

Any non-continuous construction process would be too slow and expensive to ever be realistic. Therefore, a taper on an SE is probably never going to be practical in the real world.

5. Assumption 4: Space Elevators will have climbers that go up and down the cable.

An SE could be made with moving cables to carry cargo up and down, just like an elevator in a hotel. Elevator cables provide an almost 100% efficient means of energy transfer, and the 100 MW motors that would drive them are about 99% efficient.

An SE with moving cables could have ten times more cargo throughput than SE climbers ever could. That higher throughput is because the 100 MW motors on the ground could lift the cargo five times faster than a self-contained climber could. In addition, the climber would have at least 50% of its weight as motors and power conversion equipment, whereas the moving cable car could be almost 100% cargo, providing twice the cargo in each load.

The energy cost for lifting cargo on a moving cable SE vs. climbers would be at least 6X lower. Climbers have the burden of the weight of the climber equipment, plus it is doubtful that the total power beaming efficiency for climbers could ever exceed 30%.

With a climber there are five different efficiencies that must be multiplied together to get the final efficiency. These are:

1. Laser efficiency
2. Photovoltaic efficiency
3. Climber motor efficiency
4. Atmospheric transmission efficiency
5. Tracking efficiency of the laser beam

Even if each one of these efficiencies were 80% (which is very optimistic) the total efficiency of the energy transmission would be less than 1/3.

Therefore with a moving cable SE, the energy efficiency for lifting each car would be at least three times as great. Plus each car could carry twice the load, resulting in at least six times the energy savings.

With a moving cable SE, the energy put into raising an elevator car to GEO would be recouped as the car came back down. The 100 MW motors would double as generators to recoup this energy. Only the energy required to lift the actual cargo would be expended.

Therefore, passenger travel to space would require almost zero net energy, as the energy input to send them up would be recouped as they came back down.

On the other hand, climbers do not recoup *any* of the energy that is expended in sending them to space. It is all lost. The economic and technical penalties from using climbers on an SE are so great that it is highly unlikely that climbers will ever be used in the real world.

6.   <u>**Assumption 5: In order to avoid storms and lightning, it has been assumed that an SE base station needs to be on a ship in a storm-free location.**</u>

The western equatorial Pacific Ocean region is one of the most storm-free and lightning-free locations in the world. It has been assumed that this would be a great location for an SE positioned on a ship. If a rare storm did come along, the ship could simply move out of its way.

We have shown in the discussions above relative to *Assumption 2*, that vibrational forces due to storms are not going to be a problem, but there is still the problem of lightning. The tremendous energy and heat involved in lightning could destroy any SE.

However, recent research has shown that lightning can be triggered by laser pulses. Although a number of groups had tried unsuccessfully in the past to trigger lightning strikes on demand using lasers, new research shows that a sequence of pulses is the key [8]. The methodology and equipment to be able to control and discharge lightning on demand appear to be just around the corner.

With such lightning control technology, an SE could be located in any near-equatorial location in the world. For example, Singapore, a city with one of the highest lightning-strike rates in the world, could have an SE without fear of damage.

High powered laser equipment could be located in a circle around the SE and, when a storm approached, the area could be continuously discharged to protect the SE. There would be no fear of lightning damage.

7.   <u>**Assumption 6: An SE needs to be on a mobile base in order to avoid space junk.**</u>

Space junk is an ever increasing problem, particularly in low earth orbit. In addition there are thousands of legitimate satellites that would cross the path of a SE. Because of this, there has to be a means of moving the SE out of the path of these objects.

It has been assumed that this movement would be accomplished by means of a mobile base that can move the SE to avoid collisions. However, without the need for a ship as a base to avoid storms, perhaps we need to reevaluate the need for any kind of mobile base.

With a moving cable SE, the position of the cables can be controlled using the Coriolis force on those cables. By speeding up or slowing down the cables, the Coriolis force would move the cables in or out to avoid any collision.

However, the Coriolis force movement would only operate in a general east-west direction. Any satellite or space junk that was in a purely equatorial orbit could not be avoided by east-west maneuvering.

In order to avoid any purely equatorial orbits, the SE would have to be positioned slightly off-equator. That way, any purely equatorial orbits would simply pass by the SE with no maneuvering needed.

Therefore, the SE could be constructed in a fixed location without a significant mobile base. However, just for additional security, the SE base station could also be on a short north-south track. Yet even with that, the vast majority of the debris avoidance maneuvers would still be via the Coriolis force on the cables.

8. Assumption 7: An SE will be very expensive to construct.

Huge numbers have been thrown out to predict what it will cost to eventually build an SE. Some have assumed a likely cost of $20 billion or even higher. However, a properly designed SE could be fully constructed, deployed, and put in operation for a cost of less than one billion US dollars.

In order to be as inexpensive as possible, the SE needs to be able to be constructed with a single rocket launch, be fully buildable from the ground, and have a simple, but functional base station. The methodology of such a design was given at the 2009 Space Elevator Conference [9].

The costs for constructing a fully operational SE could be as follows:

Table 1.

1)	**Materials**	$300M
	This is assuming 3M kg of tether materials at $100/kg. Although the cost of nanotubes and other strong materials is currently very high, when 3 million kilograms of anything are made in high production quantities, the cost goes way down. $100/kg may even be way high. After all, the base ingredient of the material will likely be carbon, which can be purchased, in the form of coal, for $0.10/kg. It just needs to be rearranged.	
2)	**Rocket launch to GEO**	$150M
	Although less expensive launches are available, we will assume the worst for the only launch needed.	
3)	**Initial seed satellite**	$100M
	The initial satellite to construct the SE could be remarkably simple, and need not be very expensive.	
4)	**Base station**	$100M
	Although a base station complex complete with luxury hotels and a whole new infrastructure would obviously cost more, $100M could make a functional base station in a location that needed no infrastructure upgrades. This is assuming an SE design that requires no climbers or power beaming.	
5)	**Labor and overhead**	$100M
6)	**Elevator cars**	$100M
	This would cover the design and construction of a small number of elevator cars	
7)	**Additional design and engineering costs**	$100M
	TOTAL COST:	**$950M**

9. Conclusion

After reviewing all of the above assumptions, it becomes apparent that a space elevator may not be as complicated or difficult as some had imagined. The capability to build a practical space elevator may not be that far away.

References

[1] Lee, C. et al. "Measurement of the Elastic Properties and Intrinsic Strength of Monolayer Graphene". Science 321 (5887): 385. (2008).

[2]   Banhart, F., and Ajayan, P. M.. "Carbon onions as nanoscopic pressure cells for diamond formation". Nature 382:433-43. (1996).

[3]   Belenkov, E. A. , Mavrinsky, V. , "Crystal structure of a perfect carbine", Crystallography Reports, pp 83-87, (Jan. 2008).

[4]   Itzhaki, L. et al. "Harder than Diamond: Determining the Cross-Sectional Area and Young's Modulus of Molecular Rods" Angew. Chem. Int. Ed. 44 7432-7435, (2005)

[5]   Pearson, J., "The Orbital Tower: A Spacecraft Launcher Using the Earth's Rotational Energy", Acta Astronautica 2, 785-799, (1975)

[6]   Hinton, G., "The Construction and Operation of a Practical Space Elevator Using Moving Cables", Space Elevator Conference, 2009.

[7]   Hinton, Ibid.

[8]   Kasparian, J., et al., " Progress towards lightning control using lasers" Journal of the European Optical Society - Rapid publications, Vol 3 (2008).

[9]   Hinton, Ibid.

Editors Note: This Paper was awarded an "Honorable Mention" in the 2010 Artsutanov Prize competition.

Each year, ISEC selects a theme around which it focuses the majority of its activities. One of these activities is the ISEC Report, a detailed review of the 'state-of-the-art' or status of the theme topic.

For 2010, this theme was "Space Debris Mitigation". It is generally accepted that space debris will pose a hazard to a space elevator. Sooner or later, everything which is in orbit is going to intersect the location of a space elevator. But how often will this happen? Are some regions of the space elevator more at risk than another? Questions like these must be answered to determine a) is this a manageable problem and b) what steps must be taken to keep it a manageable problem?

The authors of this report; Peter Swan, Cathy Swan and Robert "Skip" Penny took a detailed look at this issue and the 2010 ISEC Report is the result. Chapter 3 of the report, "Probability of Impact" is reproduced here. You can purchase the entire report by visiting the store at http://www.isec.org.

3.0   Determining the Probability

The probability of collision between a space elevator and space debris requires the consideration of many variables. They range from the actual population, or density, of the space debris to the velocity difference between the debris and the space elevator, to the amount of time between collisions. NASA's Orbital Debris Program Office provided the data in altitude chunks of 20 km lengths for this analysis. Each of these represents, in the probability of collision calculation, a spherical shell with the appropriate number of debris across a 20 km length of the space elevator.

3.1   Density of Space Debris (by altitude region)

It is important to estimate the densities of known and estimated [unknown] space debris to calculate the collision risk. Figure 3.1 shows densities of space debris per unit volume by altitude and is used to calculate the probability of collision.

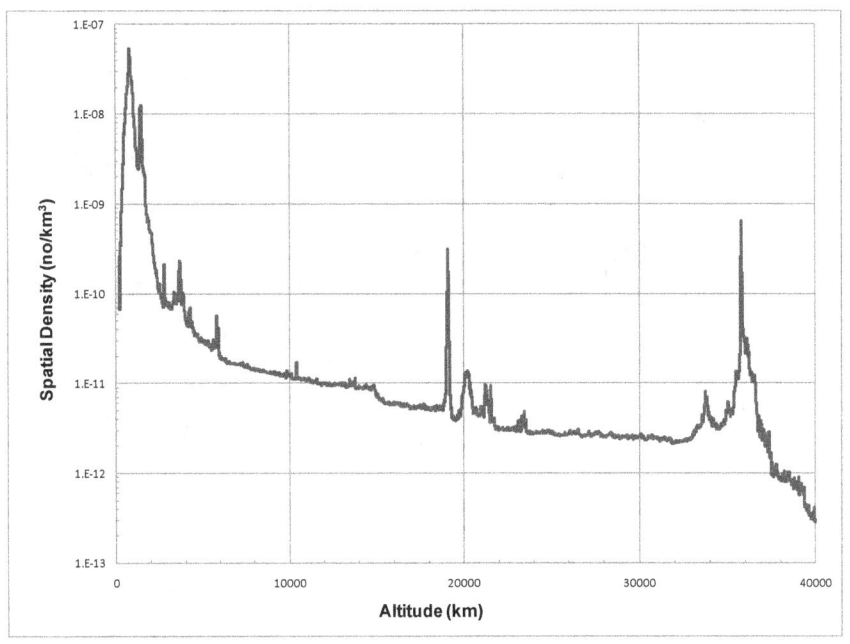

Figure 3.1 Spatial Density[1]

3.2   Relational Velocities

Determining the probability of space collisions actually requires three different calculations: head-to-head approaches, tail catch-up collisions, and oblique, or orthogonal, impacts. Each has a different set of approach velocities and areas of potential collisions. This perspective of debris to debris collision is important when considering a space elevator.   The big difference is that space elevator velocity varies linearly by altitude, not by orbital equations. At the surface of the Earth, on the equator, the

---

[1] With permission from Debra Shoots, NASA Orbital Debris Program Office, May 2010.

---

linear velocity of the anchor is 0.48 km/sec [or 360 degrees rotation in one day with the circumference of the Earth]. Each elevator segment has a unique linear velocity depending on its radius from the center of the Earth and its constant rotation. Table 3.1 shows the linear increase in velocity of a space elevator from the surface to geosynchronous altitude. The table also shows the circular orbital velocity for the appropriate altitudes. The difference is then the potential collision velocities at any given altitude. Note: the geosynchronous transfer orbit (or highly elliptical orbit) velocities are also shown at their perigees (LEO region) and apogees (GEO region).

Table 3.1: Velocity Descriptions

altitude	Circular velocity	Elliptical Velocity	Space Elevator Velocity	Elevator Impact Velocity
km	km/sec	km/sec	km/sec	km/sec
200	7.78		0.48	7.8 +/- .48
2000	6.90		0.61	6.9 +/- .6
20200	3.87		1.94	3.9 +/- 1.9
35536	3.08		3.06	+/-20's m/sec
35786	3.07		3.07	+/-10's m/sec
36036	3.07		3.09	+/-20's m/sec
600	HEO perigee	9.90	0.51	9.9 +/- .5
35786	HEO Apogee	1.64	3.07	3.1 +/- 1.6

Note: rare velocity differences at GEO could reach 125 m/sec

Table 3.2 Altitude Regions & Relational Velocities

Region	From (kilometers)	Relational Velocity (km/sec)
Super – GEO	36,036 - 100,000	Elevator Velocity + Asteroid Velocity >> 10
GEO	35,680 – 35,880	Single digits to Tens of meters/second
MEO	2,000 - 35,680 NAV – 20,200	HEO perigee = 9.9 HEO apogee = 1.6 Navigation orbit= 3.9
LEO	Spaceflight limit 200 - 2,000	Normally 6.9 to 7.4 also HEO perigee
Aero Drag	Sea Level - Spaceflight limit	Rapidly decelerating, but still significant

[GEO – geosynchronous orbit @ 35,786 km; MEO – Medium Orbit;

LEO – low Earth orbit; Radius of Earth 6378 km]

3.3   Risk of Debris to Space Elevator

Quoting from the 2001 IAA Position Paper On Orbital Debris[1], "The probability that two items will collide (PC) in orbit is a function of the spatial density (SPD) of orbiting objects in region, the average relative velocity (VR) between the objects in that region, the collision cross section (XC) of the scenario being considered, and the time (T) the object at risk is in the given region.

$$PC = 1 - e^{(-VR \times SPD \times XC \times T)}$$

This relationship is derived from the kinetic energy theory of gases which assumes that the relative motion of objects in the region being considered is random." This methodology was introduced in 1983, by Penny/Jones in their Master's thesis "A Model for Evaluation of Satellite Population Management Alternatives.[2] Note, that the PC equation may be approximated by the product of the four terms as long as the value is very small (less than 1/100). As the cataloged population, lifetime, and satellite size increase, the PC will also increase. We do not use the product method if we anticipate the PC being larger than 1/100. An example of area is (if we consider the LEO area [200 to 2,000 km altitude] of the ribbon) the cross sectional area 1,800,000 meters times 1 meter or 1,800,000 square meters, or 1.8 square kilometers. The relative velocity is the average velocity for the orbiting objects. In LEO, there are tens of thousands of tracked objects, so the calculation leads to valid estimates.

3.4   Probability of Collision (PC)

The probability of collision can be broken into separate illustrative cases. This pamphlet sets up the representation of several cases by altitude region [LEO cases A, B, & C; MEO case D; GEO case E] as identified in altitude density shells. In the LEO orbital region, two shells are 60 km in thickness and represent the area where the tracked space debris is most dense [Case A] and average [Case B]. In addition, a third case in LEO deals with all the debris from 200 to 2000 km altitude [Case C]. Another dimension for the description of LEO cases is the "untracked" (estimated) density [Cases A-u, B-u, C-u] where the numbers are estimated to be ten times the tracked numbers inside each case. A third dimension is the representation of operational spacecraft which can maneuver as they are still being controlled by the ground [Cases A-c, B-c, and C-c]. Operational spacecraft numbers are assumed to be six percent (0.06) of the tracked space debris. Case D represents MEO while Case E represents GEO. The cases are shown following:

---

[1] *2001 Position Paper On Orbital Debris, International Academy of Astronautics, 24.11.2000.*

[2] Penny Robert and Jones, Richard, "A Model for Evaluation of Satellite Population Management Alternatives," AFIT Master's Thesis, 1983.

Case A: 60 km ribbon segment (740-800 km altitude) representing the peak debris density – highest risk case.
Case B: 60 km ribbon segment (1340-1400 km altitude) representing an average debris density in LEO.
Case C: 1800 km ribbon segment (200-2000 km altitude) representing the entire LEO environment.
Case A-u, B-u, C-u: represent the untracked items in above described segments. Estimated to be ten times the tracked debris.
Case A-c, B-c, C-c: represent the controlled satellites in above segments. Estimated to be six percent of the tracked debris.

Medium Earth Orbit (1 case)

Case D: 200 km ribbon segment (around 20,200 km altitude) representing the navigation orbit environment [only tracked items are calculated].

GEO Orbit (1 case)

Case E: 200 km ribbon segment (35,680 - 35,880 km altitude) representing the GEO environment [only tracked items are calculated].

As we noted earlier, the probability of collision is a function of the relative velocity (VR), the density of objects (SPD), the cross sectional area (XC) and time (T). This approach works well for LEO where the behavior of Earth orbiting objects is very similar to the behavior of gas molecules (as noted in Section 3.1). It is less similar for MEO and GEO; however, we use the same methodology as we lack anything better.   We will use the formula $PC = 1 - e^{(-VR \times SPD \times XC \times T)}$ for all eleven cases.

3.4.1. LEO Cases

The three baseline cases for LEO tracked debris will be run for the probability of collision (PC) in LEO for three threat types: untracked (< 10 cm), tracked (> 10 cm), and cooperative satellites. This range of altitude segments and debris types attempts to layout the range of threats that a space elevator will encounter in day to day operations in Low Earth Orbit. This requires a total of nine cases for LEO predictions of collision.

Table 3.3 LEO Regional Breakout by Cases

Types of Debris	Case	Comment
Untracked Debris < 10 cm		10 x tracked
60 km stretch - peak	A-u	Highest Density
60 km stretch - average	B-u	Average LEO
LEO 200 - 2000 km	C-u	Total LEO region
Tracked Debris > 10 cm		
60 km stretch - peak	A	Highest Density
60 km stretch - average	B	Average LEO
LEO 200 - 2000 km	C	Total LEO region
Cooperative Debris		0.06 x tracked
60 km stretch - peak	A-c	Highest Density
60 km stretch - average	B-c	Average LEO
LEO 200 - 2000 km	C-c	Total LEO region

The numbers of objects tracked in LEO are illustrated by Figure 3.2 from NASA's Orbital Debris Program Office. As seen, there is a peak at 740-800 km. In addition, if you do the "eye-ball" smoothing across 200-2000 km, 1340-1400 km reflects an average density. As reflected in this chart, two-thirds of all tracked debris are between 200 and 2000 km in altitude. The following table (Table 3.4) shows the significant case of tracked space debris with the calculated probability of collisions between LEO tracked debris and a space elevator.

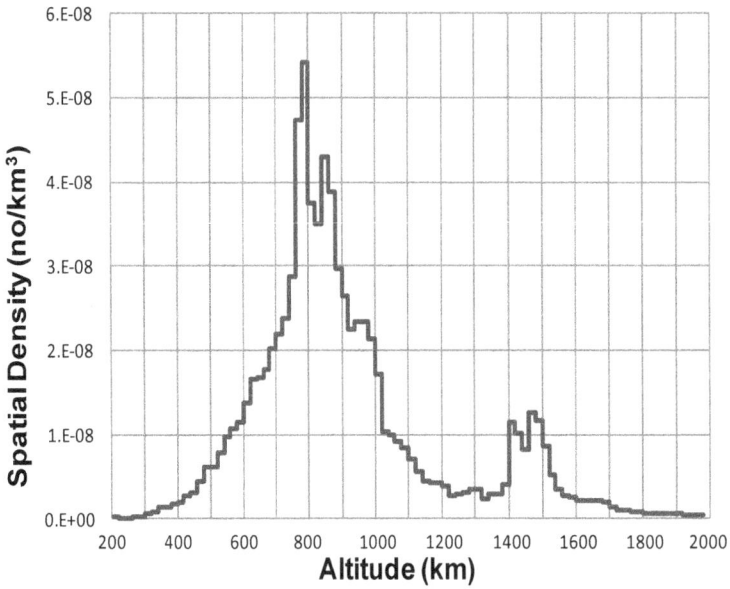

Figure 3.2 LEO Spatial Density[1]

---

[1] With permission from Debra Shoots, NASA Orbital Debris Program Office, May 2010.

Table 3.4 Probability of Collision for Tracked Objects

	PC Case A 60 km all tracked, (peak band)	PC Case B 60 km all tracked, (LEO avg)	PC Case C LEO all tracked objects
Top Altitude	800	1400	2000
Bottom Altitude	740	1340	200
Tracked Objects	1672	149	11298
Ribbon Area (km^2)	.06	.06	1.8
Time (days)	365.25	365.25	365.25 (1)
Probability of Collision	**0.457859 per year**	**0.043647 per year**	**0.969317 per year (.00949 per day)**

- **Case A** Results show that the tracked items have a one-in-two chance of having a conjunction with the space elevator each year across a 60 km segment in the high threat region. Very limited number of space elevator 60 km segments across LEO are at high density risk levels.
- **Case B** Shows that the average in LEO, for any 60 km segment, is around one-in-twenty chances per year [most LEO 60 km segments have less].
- **Case C** The full spread across LEO shows the probability of conjunction for tracked objects is essentially three per year. This means that some location across 1,800 km will have a potential conjunction by tracked debris every four months.

One must remember that, when dealing with tracked objects, we know where the debris is and can predict its future location to enough precision to enable us to make judgments as to the specific risk per opportunity for conjunction.

Similar calculations were conducted across three cases for the "small stuff," or untracked debris. The summary probability of collision for a space elevator with untracked objects is shown in Table 3.5.

Table 3.5 Probability of Collision for Untracked Objects

	PC Case A-u 60 km all un-tracked, (peak band)	PC Case B-u 60 km all un-tracked, (LEO avg)	PC Case C-u LEO all un-tracked objects
Top Altitude	800	1400	2000
Bottom Altitude	740	1340	200
Tracked Objects	16720	1490	112980
Ribbon Area (km²)	.06	.06	1.8
Time (days)	365.25 (1)	365.25 (1)	365.25 (1)
Probability of Collision	**0.9978 per year (.0166 per day)**	**0.3500 per year (.00122 per day)**	**0.9999999 per year (.0949 per day)**

This second type of debris is the untracked set, which was earlier estimated to be roughly ten times the density of the tracked set. With this as the starting position, the probability of conjunction (PC) for:

Case A-u is one-in-60 days,

Case B-u is one-in-700 days, and

Case C-u is one-in-ten days.

The cross-sectional area of the untracked space debris is less than 10 cm (with the preponderance much smaller) which, because of its velocity difference, should just "blow through" the ribbon when it actually collides.

To put this in perspective, if you were to look at the probability of collision for one square meter of space elevator ribbon (1 m wide by 1 m long) in Low Earth Orbit (200-2000 km altitude), the probability is about once every 2,000 years for any specific ribbon square meter. As the danger area is the full LEO environment (200-2000 km length), the summation of these probabilities for each of the 1,800,000 meter squares is equivalent to once every ten days. However, the probabilities of multiple impacts on any single square meter of ribbon are extremely small!

The third type of debris is the tracked and cooperative set, which includes all operational satellites in Low Earth Orbit. Space elevator operators will track operational satellites; and, then, work with the owner as to appropriate actions to ensure collision avoidance. This is beneficial to both parties. As this is approximately 6 % of the tracked debris, the probabilities are as follows:

Case A-c yields a collision every 30 years

Case B-c yields a collision every 400 years

Case C-c yields a collision every 5 years

The summary of the probability of collision for space debris with a space elevator in the LEO region is summarized below:

Table 3.6 Probability of Collision LEO Summary

Types of Debris	Case	Probability of Collision
Untracked Debris < 10 cm		PC per day
60 km stretch - peak	**A-u**	1.66%
60 km stretch - average	**B-u**	0.12%
LEO 200 - 2000 km	**C-u**	9.54%
Tracked Debris > 10 cm		PC per year
60 km stretch - peak	**A**	45.79%
60 km stretch - average	**B**	4.36%
LEO 200 - 2000 km	**C**	96.93%
Cooperative Debris		PC per year
60 km stretch - peak	**A-c**	Every 30 yrs
60 km stretch - average	**B-c**	Every 400 yrs
LEO 200 - 2000 km	**C-c**	Every 5 yrs

3.4.2. Medium Earth Orbit Case

The medium altitude orbit covers a range of mostly empty space. The region around a 12 hour circular orbit for navigation satellites is populated (estimate 200+ satellites) with circular orbits of eight to twenty satellites per orbital plane; however, the good news is the volume is huge. The spherical shell (200 km in radius height) centered at 20,200 km altitude is labeled as case D. Another case is the geosynchronous transfer orbit with rocket bodies and satellite residuals. This orbit has a lot of residual rocket bodies; but, they are numerically not a large threat because of the vast volume and the location of perigee. This case is described as:

Case D: 200 km ribbon segment (20,200 km altitude) representing the Navigation orbit environment. (see Table 3.7 for results)

3.4.3. GEO Case

The GEO belt is extremely interesting and has many operational spacecraft generating large profits for commercial enterprises. As such, the space elevator must not interfere with GEO operational satellites. In addition, derelict spacecraft in this orbit are all going in the same direction as the space elevator. This means that the likelihood of fragmentation of these satellites or damage to the tether is greatly reduced. This case is described as:

Case E: 200 km ribbon segment (35,680 - 35,880 km altitude) representing the GEO environment. (see Table 3.7 for results)

Table 3.7 Probability of Collision for Tracked Objects

	PC Case D 200 km all tracked, at MEO	**PC Case E** 200 km all tracked, at GEO
Top Altitude	20300	35880
Bottom Altitude	20100	35680
Tracked Objects	22	600
Ribbon Area (km^2)	0.2	0.2
Time (days)	365.25	365.25
Probability of Collision	**0.00030 per year**	**0.0026 per year**

The probabilities of collision are as follows:

Case D is 3 in 10,000 years

Case E is 3 in 1,000 years

3.5   Summary Probability of Collision

After evaluating all of the eleven cases, the numbers show that LEO is the highest threat arena. We know this intuitively as the density of space debris is greatest at LEO and it has the highest differential velocities – two major drivers in the probability of collision equation. In addition, as the population density is not great at MEO and the volume is huge, MEO still falls into the "Big Sky Theory" of less worrisome. The GEO orbit has a restrictive band [Sir Arthur Clarke's altitude for station keeping at zero latitude] of limited population. This leads to some concern from the numbers; however, the differences in velocities are so small that the danger is even smaller. The next chart summarizes the concerns for all eleven cases.

Table 3.8 Summary of Probability of Collisions

Types of Debris	Case	Collision About Every
Untracked <10 cm		
60 km stretch peak	A-u	60 days
60 km stretch average	B-u	2.5 years
LEO 200-2000 km	C-u	10 days
Tracked Debris >10cm		
60 km stretch peak	A	2 years
60 km stretch average	B	23 years
LEO 200-2000 km	C	1.3 years
Cooperative Objects		
60 km stretch peak	A-c	30 years
60 km stretch average	B-c	400 years
LEO 200-2000 km	C-c	5 years
Tracked Debris >10cm		
200 km stretch-MEO	D	> 4000 years
200 km stretch-GEO	E	400 years

These results lead us to the following conclusions:

GEO altitude belt is not a problem.

MEO volume is not a problem.

Untracked, small (<10 cm debris) will, on the average, impact a Space Elevator in LEO (200-2000 km) once every ten days, and therefore, the ribbon must be designed for impact velocities and energies.

Tracked debris will impact the total LEO segment (200 – 2000 km) once per 100 days or multiple times a year if no action is taken.

Tracked debris will, on average, impact a single 60 km stretch of LEO space elevator every 18 years and every five years in the peak regions if no action is taken.

(end of excerpt)

Each year, as part of ISEC's theme activities, ISEC produces a poster reflecting that year's theme. These posters are shown on the following pages.

- For 2009, as this was ISEC's first year, there was no theme. Instead, we chose to commemorate the Space Elevator Games (organized by the Spaceward foundation – http://www.spaceward.org with prize funding provided by NASA - www.nasa.gov/challenges/) as 2009 was the first year that a winner was awarded prize money. LaserMotive (www.lasermotive.com), out of Seattle, Washington, won $900,000 for their entry's performance. The Kansas City Space Pirates and the USST (University of Saskatchewan Space Design Team) teams are also pictured.

- For 2010, the theme was SPACE DEBRIS MITIGATION – SPACE ELEVATOR SURVIVABILITY. The poster shows two pictures. The topmost one shows a Repair Climber travelling up/down the tether, making small repairs as it detects holes in the tether caused by small pieces of space debris. The bottom picture shows a Climber using its onboard engine to induce an oscillation in the tether – this to move it out of the way of satellites and larger space debris. A combination of these techniques, plus periodic replacement of tether segments, should keep the ribbon robust and able to carry cargo.

- For 2011, the theme was RESEARCH AND THOUGHT TARGETED TOWARDS THE GOAL OF A 30 MYURI TETHER. A Yuri (named in honor of Yuri Artsutanov) is a unit of Specific Strength. It is equivalent to 1 Pascal-cubic-meter per kilogram. A Mega Yuri (MYuri) is equivalent to the commonly used units of 1 GigaPascal-cubic-centimeter per gram (1 GPa-cc/g) and to 1 Newton per Tex (N/Tex). Current thinking has a tether with a strength of 30 MYuri as being strong enough to build an earth-based space elevator.

These posters are 11x17 inches in size, are in full color, and are offset-printed on heavy-duty, glossy stock paper.

ISEC Members automatically receive that year's poster for each year they are a member in good standing. If you are not a member, and wish to purchase any of these posters, please visit the store on our website http://www.isec.org.

All of these posters were designed by Mr. Frank Chase and we at ISEC are very grateful for his artistic contributions towards our goal of building a Space Elevator. Frank can be reached at kahuna.frank@gmail.com.

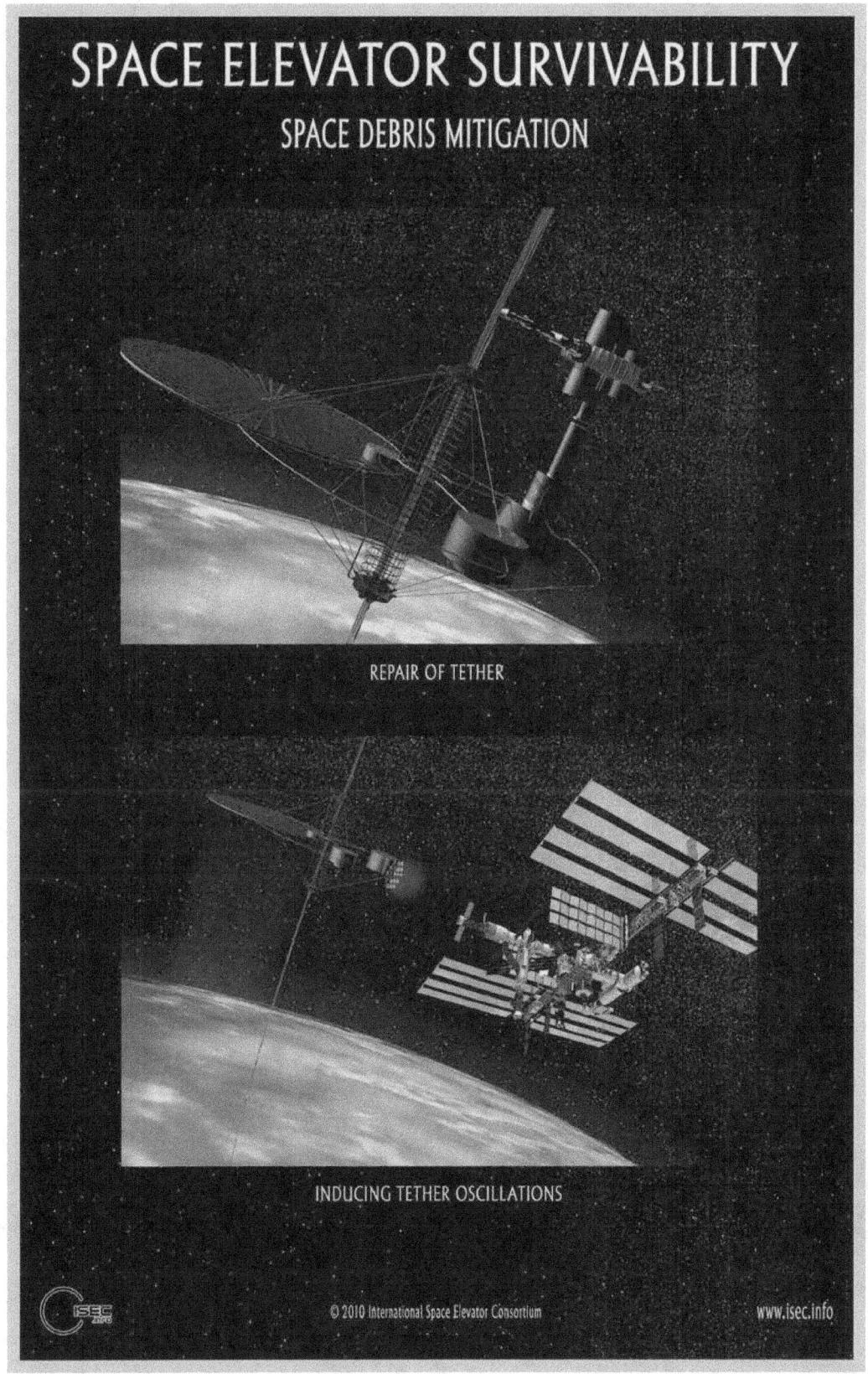

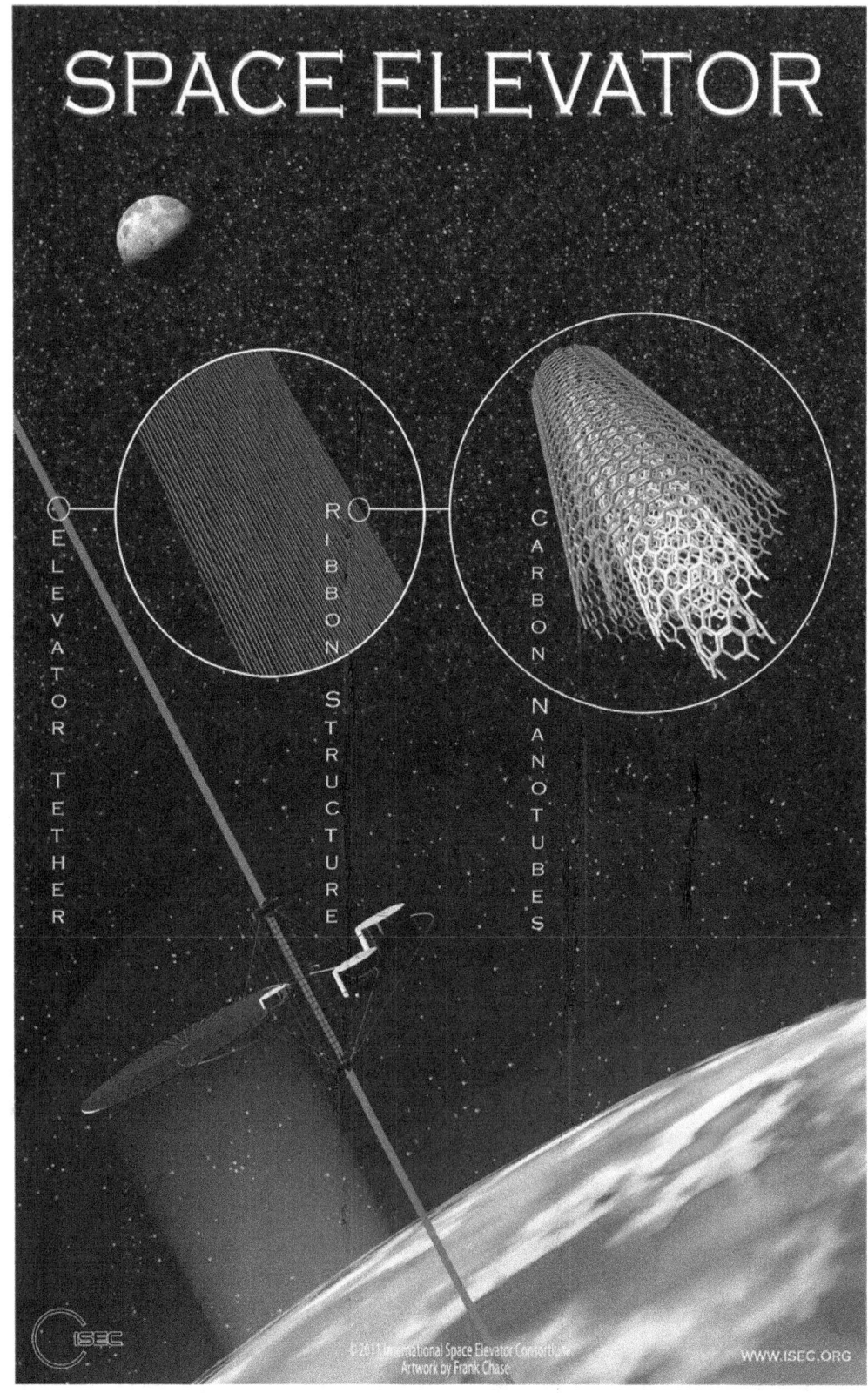